NEEM FOR THE MANAGEMENT OF CROP DISEASES

NEEM FOR THE MANAGEMENT OF CROP DISEASES

–Editor–
V. Mariappan
Professor & Head, Department of Plant Pathology
Tamil Nadu Agricultural University
Coimbatore-641 003 (India)

2016
Associated Publishing Company
A Division of
Astral International Pvt. Ltd.
New Delhi - 110 002

First Published, 1995
Reprinted, 2016

ISBN: 978-93-5130-984-0 (International Edition)

Published by : **Associated Publishing Company**
A Division of
Astral International Pvt. Ltd.
– ISO 9001:2008 Certified Company –
4760-61/23, Ansari Road, Darya Ganj
New Delhi-110 002
Ph. 011-43549197, 23278134
E-mail: info@astralint.com
Website: www.astralint.com

Typesetting : **Divya Computers, Meerut-250 002**

Digitally Printed at : **Replika Press Pvt. Ltd.**

CONTENTS

FOREWORD

Neem *Azadirachta indica* has been known for its medicinal and pesticidal properties and the farmers of our country used the neem products for the management of diseases and pests of crop plants from ancient time. However from the dawn of this century, broad spectrum synthetic chemical pesticides have relegated the neem and other plant products for various other miscellaneous purposes. The indiscriminate and large scale use of highly poisonous synthetic chemical pesticides have resulted into ecological imbalances besides their toxic effect on all living organisms including human population. Hence there is an urgent need for developing methods and materials with different standards which could be ecofriendly. Among the several materials available neem products have been reported to be a potential pesticide in recent years. The usefulness of neem in the management of crop diseases caused by pathogens like fungi, viruses and nematodes have been well demonstrated in the recent years.

The present book "Neem for the Management of Diseases of Crop Plants" provide evidence and justification for using neem products for the management of diseases affecting cereals, pulses, oil seeds and horticultural crops.

I hope the information contained in this book would be of immense value and guidance to the scientists, students, farmers as well as to those concerned with ecological situations and organic farming.

Tamil Nadu
Agricultural University
Coimbatore

S. SANKARAN
Vice Chancellor

PREFACE

Neem (*Azadirachta indica*) has emerged as a single plant species possessing potential pesticidal properties among the several plant species known to mankind. In ancient periods neem and other plants were used extensively for medicinal and pesticidal properties. However with the advent of broad spectrum, highly poisonous synthetic pesticides, the plant products have been relegated to other miscellaneous purposes. The chemical pesticides have been highly useful in keeping the insect pests and diseases caused by pathogens under check and helped to increase the yield to a phenomenal extent during post war period. But their extensive and indiscriminate uses have resulted into several problems like, toxicity to all living organisms, including man, destruction of beneficial organisms, development of resistance by insect and pathogens etc. Thus the ecological hazards caused by the chemical pesticides are posing a great threat to the biosphere. The realisation of the harmful effects caused by the chemical synthetic pesticides have forced the scientists and farmers to search for the alternate materials so as to avoid the ecological hazards. Among the several materials available plant products offered greater scope than the rest as they are relatively safe, easily biodegradable and ecofriendly. Neem (*A. indica*) among the many plant species has been reported to possess insecticidal, antifungal, antibacterial, antiviral properties. During the past three decades innumerable number of publications have appeared through various forums and most of them were on its efficacy in the control of insect pests. Relatively less number only appeared on its efficacy on the pathogenic populations (fungi, bacterial viruses) and nematodes of crop plants. The informations available from the research efforts of earlier decades were mostly from the exploratory investigations. The usefulness of neem and other plant products for the management of crop diseases of food crops and commercial crops have been demonstrated by many scientists all over the country and are available in recent years. Hence an attempt has been made to compile these informations and thus the book is unique in presenting the recent data on the evidences of management of crop diseases by using neem products. I hope the contents of this book is useful to the readers, scientists and farmers for utilizing the neem products as a component in the Integrated Pest Management for the management of crop diseases.

The versatile nature of neem have been well demonstrated by many eminent scientists like Dr. H. Schumetterer, Dr. R.C. Saxena and others in the past decades. I owe my gratitude to Dr. R.C. Saxena former Entomologist of International Rice Research Institute, Manila, Philippines who has been the mentor of my devotion to neem research.

V. MARIAPPAN

A

Management of Diseases Caused by Fungi and Bacteria

Neem for the Management of Crop Diseases, 1995
Editor : V. Mariappan
Associated Publishing Co., New Delhi, India
pp. 3-10.

MANAGEMENT OF RICE BLAST *PYRICULARIA ORYZAE* BY USING NEEM (*AZADIRACHTA INDICA*) AND OTHER PLANT PRODUCTS

Mariappan, V., Rajeswari, E. and Kamalakannan, A.

Department of Plant Pathology, Tamil Nadu Agricultural University
Coimbatore-641 003, Tamil Nadu

Rice is one of the important staple food grain for mankind and is grown in many countries. In India rice is grown in 42.2 million hectares with an annual production of 110 million tonnes (IRRI, 1994). The average rice production in India is about 2.6 t/ha which is far behind the yield of 5.7 t/h in China. Several reasons including abiotic and biotic stresses have been attributed for the low yield obtained in our country. Among the biotic factors, damages caused by insects and diseases are more devastating and the blast disease caused by *Pyricularia oryzae* is the most serious disease among the several diseases that affect the rice crop. Rangaswami and Subramanian (1957) reported a loss in yield upto 70 per cent under severe incidence of blast disease in Tamil Nadu. Padmanabhan (1963) estimated that rice production was reduced by 9.8 per cent in the state of Jammu and Kashmir due to blast disease. Losses upto 30-100 per cent have been reported from Guyana (Bisessar, 1965). In U.S.A. where the overall loss from diseases of rice has been estimated at six per cent of the crop yield, 22 per cent of the amount has been attributed to blast disease alone (Cramer, 1967). Several methods have been used for the control of the disease and among them chemical control has been the most widely adopted practice in many countries. The ecological concern over the excess use of pesticides and the spiralling up cost of the chemical pesticides have rendered the farmers to select out methods that are environment friendly and also relatively cheap in cost. The last two decades had witnessed an increased interest on investigations on plants as a source of pest control material and as a possible alternative to synthetic pesticides.

The neem (*Azadirachta indica*) and other plant species have been reported to possess antimycotic properties against *P. oryzae.* Lapis and Dumancas (1979) screened 93 plant species and reported 19 of them were inhibitory to *P. oryzae.* Krishnamohan *et al.* (1988) stated that leaf extracts of onion (*Allium cepa*), garlic (*A. sativum*) and neem (*A. indica*) were effective in reducing the mycelial growth of *P. oryzae.* Many plant species were reported exert an adverse effect on the growth of *P. oryzae* (Garcia *et al.,* 1991, Tewari and Premalatha Dath, 1984, Muthusamy *et al.,* 1988, Miah *et al.,* 1990, Mishra and Tewari, 1990; Tewari and Mandakini Nayak, 1991; Mishra *et al.,* 1992. Similarly extracts of leaves of some plants species have been reported to inhibit the germination of *P. oryzae conidia* (Mishra *et al.,* 1992).

As the plant products were reported to be effective against the rice blast pathogen (*P. oryzae*) attempt was made to screen the plants available around Coimbatore for their efficacy so as to select out the best material for field application. The leaf extracts were prepared at 10 per cent concentration by pounding the leaves with appropriate quantity of water and the filtered extract was used in the experiment. These preparations were used at 500 l/ha for spraying. For testing the efficacy of the preparations on growth and germination of *P. oryzae* crude extract of the leaves was mixed with the media in appropriate quantity to obtain the required concentration.

Effect of leaf extracts of plants on the growth of *P. oryzae*

The leaf extract of 21 plant species were added to the potato dextrose agar medium individually and inoculated with the eight day old culture of *P. oryzae* at the centre of the petriplate. For control the fungus culture was inoculated to the potato dextrose agar medium. Ten days after inoculation the growth (diameter of the colony) was measured and the percentage of inhibition was worked out. From the results in (Table 1) it may be seen that the leaf extracts of *Prosophis juliflora,*

Table 1. Effect of leaf extract (cold water) on the growth of *P. oryzae*

S. No.	Plant Species	Inhibition per cent over control
1.	*Adenanthera pavonia*	16.00 lk
2.	*Adhatoda vasica*	87.40 b
3.	*Aegle marmelos*	35.00 h
4.	*Antigonon leptopus*	4.76 on
5.	*Azadirachta indica*	66.50 d
6.	*Bougainvillaea spectabilis*	20.23 ij
7.	*Caesalpinia pulcherrima*	6.13 n
8.	*Catharanthus roseus*	46.13 f
9.	*Centella asiatica*	11.37 m
10.	*Datura metel*	5.47 n
11.	*Eucalyptus globulus*	18.57 jk
12.	*Ipomea carnea*	46.80 f
13.	*Lawsonia inermis*	48.10 f
14.	*Melia azadirach*	5.90 n
15.	*Neerium indicum*	14.67 i
16.	*Ocimum sanctum*	40.26 g
17.	*Phyllanthus fraternus*	53.33 e
18.	*Polyalthia longifolia*	64.50 d
19.	*Prosophis juliflora*	93.33 a
20.	*Solanum nigrum*	4.17 on
21.	*Solanum tarvum*	3.50 o
22.	*Tamarindus indicus*	23.83 i
23.	*Vitex negundo*	72.40 c
	Mean	34.29

All values in percentages. Mean of 3 replications In a column means followed by samé letter(s) are not significantly different (P = 0.05) by DMRT

Adhatoda vasica, Vitex negundo and *A. indica* at 10 per cent concentration inhibited the growth significantly to an extent of 93.3, 87.4, 72.4 and 66.50 per cent respectively over control.

In another experiment the leaf extracts of 20 plant species were tested and among them *P. juliflora, Zizyphus jujuba, Abutilon indicum, Cynodon dactylon, Cyperus rotundus, Clerodendron inerme, Lantana camara, Boerhaavia diffusa* and *Azadirachta indica* inhibited of the growth of *P. oryzae* from 93.3 to 66.8 per cent over control (Table 2).

Table 2. Effect of leaf extract (cold water) on the growth of *P. oryzae*

S. No.	Plant Species*	Inhibition per cent over control
1.	*Abutilon indicum*	78.1 c
2.	*Acalypha indica*	64.1 j
3.	*Argemone mexicana*	47.7 o
4.	*Aristolochia bracteata*	51.6 n
5.	*Azadirachta indica*	66.8 i
6.	*Bauhinia purpurea*	28.4 s
7.	*Boerhaavia diffusa*	67.8 h
8.	*Chenopodium album*	33.6 r
9.	*Clerodendron inerme*	72.9 f
10.	*Cynodon dactylon*	77.8 d
11.	*Cyperus rotundus*	74.1 e
12.	*Eclipta alba*	22.3 t
13.	*Erythrina indica*	42.4 p
14.	*Lantana camara*	69.3 g
15.	*Mirabilis jalapa*	57.3 l
16.	*Morinda tinctoria*	34.7 q
17.	*Portulaca oleraceae*	56.8 m
18.	*Prosophis juliflora*	93.3 a
19.	*Tribulus terrestris*	59.3 k
20.	*Zizyphus jujuba*	82.0 b

* Plant extracts at 10 per cent concentration
All values in percentages. Means of 3 replication. In a column means followed by same letter(s) are not significantly different (P = 0.05) by DMRT.

The biomass production of *P. oryzae* when grown in Richard's broth containing the leaf extracts of 23 plant species at 10 per cent concentration was inhibited considerably. The leaf extracts of *P. juliflora, A. vasica, V. negundo* and *A. indica* were significantly effective and reduced the biomass production of *P. oryzae* to an extent of 100.0, 89.70, 71.4 and 60.8 per cent respectively over control (Table 3).

The germination of the spores of *P. oryzae* has been found inhibited by the leaf extracts of 23 plant species (in cold water) when mixed with the spore suspension (Table 4). Among them the leaf extracts of *P. juliflora, A. vasica, V. negundo* and *A. indica* were significantly effective and recorded an inhibition of 80.47, 76.73, 59.37 and 55.60 per cent over control respectively.

Table 3. Effect of leaf extract on the biomass production of *P. oryzae*

S. No.	Plant Species	Percent inhibition of biomass production
1.	*Adenanthera pavonia*	16.73 k
2.	*Adhatoda vasica*	89.70 b
3.	*Antigonon leptopus*	4.10 p
4.	*Azadirachta indica*	60.83 d
5.	*Aegle marmelos*	42.50 h
6.	*Bougainvillaea spectabilis*	15.17 l
7.	*Caesalpinia pulcherrima*	5.67 o
8.	*Catharanthus roseus*	45.93 g
9.	*Centella asiatica*	5.77 o
10.	*Datura metel*	6.00 o
11.	*Eucalyptus globulus*	8.97 m
12.	*Ipomea cornea*	41.57 h
13.	*Lawsonia inermis*	47.33 g
14.	*Melia azadirach*	7.47 h
15.	*Neerium indicum*	21:23 j
16.	*Ocimum sanctum*	51.93 f
17.	*Phyllanthus fraternus*	46.60 g
18.	*Polyalthia longifolia*	55.17 e
19.	*Prosphis juliflora*	100.00 a
20.	*Solanum nigrum*	5.40 o
21.	*Solanum tarvum*	6.97 h
22.	*Tamarindus indicus*	31.00 i
23.	*Vitex negundo*	71.43
	Mean	34.28

Concentration of plant extract
All values in percentages. Mean of 3 replications In a column mean followed by same letter(s) are not significantly different (P = 0.05) by DMRT.

Table 4. Effect of cold water extract of leaves on the *P. oryzae* spore germination

S. No.	Plant Species*	Per cent inhibition of spore germination
1.	*Adenanthera pavonia*	13.73
2.	*Adhatoda vasica*	76.73
3.	*Aegle marmelos*	34.23
4.	*Antigonon leptopus*	3.46
5.	*Azadirachta indica*	55.60
6.	*Bougainvillaea spectabilis*	22.47
7.	*Caesalpinia pulcherrima*	7.67
8.	*Catharanthus roseus*	46.46
9.	*Centella asiatica*	9.60
10.	*Datura metel*	6.30
11.	*Eucalyptus globulus*	15.37
12.	*Ipomea carnea*	35.30
13.	*Lawsonia inermis*	45.17
14.	*Melia azadirach*	5.60
15.	*Neerium indicum*	15.23

S. No.	Plant Species*	Per cent inhibition of spore germination
16.	*Ocimum sanctum*	45.03
17.	*Phyllanthus fraternus*	47.27
18.	*Polyalthia longifolia*	54.57
19.	*Prosophis juliflora*	80.47
20.	*Solanum nigrum*	6.43
21.	*Solanum tarvum*	5.93
22.	*Tamarindus indicus*	20.67
23.	*Vitex negundo*	59.37
	Mean	30.99

* Plant extracts at 10 per cent concentration
Hours after incubation
All values in percentages. Mean of 3 replications In a column means followed by same letter(s) are not significantly different (P = 0.05) by DMRT.

The leaf extracts of selected five plant species obtained by using cold water, ethanol, methanol, benzene and acetone were tested for their efficacy in reducing the germination of the spores of *P. oryzae*. The percentage germination of the spores of *P. oryzae* in water (Control) was 89.9 per cent while the germination per cent in the cold water extract containing the leaf extracts of all the five plant species including neem was reduced significantly (Table 5). The leaf extracts obtained by using the solvents (ethanol, methanol, benzene and acetone) also were effective in reducing the rate of germination of the spores of *P. oryzae* and they were relatively more effective than the cold water extracts.

Table 5. Effect of leaf extract obtained by using different solvents on the spore germination of *P. oryzae*

S. No.	Plant Species*	Percentage of spore germination				
		Distilled water	Ethanol	Methanol	Benzene	Acetone
1.	*Abutilon indicum*	24.2 c	21.8 c	22.6 c	20.2 b	22.4 c
2.	*Azadirachta indica*	35.9 d	33.4 e	35.2 e	30.5 d	34.8 d
3.	*Cynodon dactylon*	37.4 e	26.4 d	30.0 d	22.4 c	32.6 d
4.	*Prosophis juliflora*	17.2 a	17.1 a	17.2 a	16.5 a	17.5 a
5.	*Zizyphus jujuba*	19.7 b	17.9 b	18.7 b	16.8 a	18.2 b
	Control	89.9				

* Plant extracts at 10 per cent concentration
All values in percentages. Mean of 4 replications
In column means followed by same letter(s) are not significantly different (P = 0.05) by DMRT.

The rice seeds (IR50 variety) soaked in leaf extracts of selected five plant species for 12 hours were sown in pots. The seedlings of 15 days old were inoculated with the spore suspension of *P. oryzae* (at 40,000 cells/cml). Control seedlings were soaked in water and inoculated with the spore suspension. The severity of the disease was scored using 0-9 scale of the standard Evaluation System (IRRI, 1988) and Percent Disease Index (PDI) was worked out by counting 25 leaves (Mckinney, 1928). The results are presented in Table 6. The reduction in the PDI values in all the treatments were significantly less than that of the control.

Table 6. Effect of seed treatment with leaf extracts on the incidence of rice blast disease

S. No.	Plant Species*	PDI-values - Percent reduction over control		
		7**	15**	25**
1.	*Abutilon indicum*	30.2 d	21.5 d	15.0 d
2.	*Azadirachta indica*	38.8 c	28.9 c	20.3 c
3.	*Cynodon dactylon*	26.8 c	18.7 e	12.8 e
4.	*Prosophis juliflora*	50.9 a	42.9 a	31.6 a
5.	*Zizyphus jujuba*	45.5 b	39.0 b	29.4 b

* Plant extracts at 10 per cent concentration
** Days after inoculation
All values in percentage. Mean of 3 replications. In column means followed by same letter(s) are not significantly different (P = 0.05) by DMRT.

The leaf extracts of ten plant species at 10 per cent concentration was sprayed on rice plants (IR 50 variety) raised in pots. Three hours later the plants were inoculated with the conidial suspension of the pathogen (*P. oryzae*). The blast lesions started appearing after 7 days. The lesions were graded and per cent disease index was worked out on 7, 15 and 25 days after inoculation. The treatments *P. juliflora, Zizypus jujuba, A. indica* have recorded a reduction in the PDI value to an extent of 54.7, 53.2 and 43.4 per cent over control (Table 7).

Table 7. Effect of pre-inoculation treatment on rice with leaf extracts on the incidence of rice blast disease

S. No.	Plant Species*	PDI-values - Percent reduction over control		
		7**	15**	25**
1.	*Abutilon indicum*	34.3 d	26.6 d	19.8 d
2.	*Acalypha indica*	21.2 i	11.3 j	7.1 j
3.	*Azadirachta indica*	43.4 c	34.7 c	28.6 c
4.	*Boerhaavia diffusa*	24.6 h	13.0 i	7.6 i
5.	*Clerodendron inerme*	29.1 f	16.5 g	11.5 g
6.	*Cynodon dactylon*	31.7 e	24.6 e	19.1 e
7.	*Cyperus rotundus*	29.2 f	20.0 f	12.5 f
8.	*Lantana camara*	26.7 g	14.6 h	8.5 h
9.	*Prosophis juliflora*	54.7 a	47.2 a	38.8 a
10.	*Zizyphus jujuba*	53.2 b	46.5 b	36.6 b

* Plant extracts at 10 per cent concentration
** Days after inoculation
All values in percentage. Mean of 3 replications. In a column means followed by same letter(s) are not significantly different (P = 0.05) by DMRT.

The leaf extracts of the plant species *Abutilon indicum, Azadirachta indica, Prosophis juliflora* and *Zizypus jujuba* were tested for their efficacy in controlling the rice blast disease in field in comparison with the standard fungicide carbendazim and the chemical nickel nitrate. The field trial was laid out in randomized block design with four replication. Rice. variety IR50 which is highly susceptible for blast was transplanted. The leaf extracts (at 10 per cent), nickel nitrate (at 10^{-4} M) and carbendazim (0.1 per cent) were sprayed at 500 l/ha at 15, 30 and 45 days after transplanting. The blast lesions started appearing 40 days after transplanting and were scored, using 0-9 scale. The per cent Disease Index (PDI), and reduction in PDI values over control treatment, was worked out for each treatment. The yield recorded and the cost benefit ratio are presented in Table 8.

Table 8. Efficacy of leaf extracts and chemicals on the control of rice blast disease under field condition

S. No.		Concentration	PDI values per cent reduction over control		Yield kg/ha	Cost benefit ratio
			45 DAT	60 DAT		
1.	*Abutilon indicum*	10%	12.2 f	8.0 f	2778.3 e	1.6
2.	*Azadirachta indica*	10%	19.4 e	13.7 e	3350.0 d	6.3
3.	*Prosophis juliflora*	10%	37.7 b	34.5 b	4025.0 b	12.3
4.	*Zizyphus jujuba*	10%	26.0 d	19.5 d	3418.7 cd	6.8
5.	Nickel nitrate	10^{-4}M	30.3 c	24.7 c	3522.3 c	4.3
6.	Carbendazim	0.1%	39.4 a	35.1 a	4165.3 a	7.0
	Control	—	—	—	2596.7 f	—

DAT Days after transplanting
Mean of 3 replications.
In a column means followed by same letter(s) are not significantly different (P = 0.05) by DMRT.

Among, the treatments the fungicide carbendazim has been significantly effective in reducing the disease severity than the other treatments. It was followed by *P. juliflora*, nickel nitrate, *Z. jujuba, A. indica* and *A. indicum* in the order of their efficacy both at 45 and 60 DAT.

Maximum yield of 4165.3 kg/ha was obtained from the treatment 'carbendazim', followed by an yield of 4025.0, 3522.3, 3418.7, 3350.7 and 2778.3 kg/ha in the treatments of *P. juliflora*, nickel nitrate, *Z. jujuba, A. indica* and *A. indicum* respectively. The yield obtained from all the treatments were significantly higher than the yield of 2596.7 kg/ha obtained from control treatment. However the cost benefit ratio obtained from *P. juliflora* was relatively higher (12.3) than those obtained from other treatments including the fungicide or chemical treatments. Application of *Z. jujuba* and *A. indica* also recorded higher cost benefit ratio and were comparable to the cost benefit ratio obtained from the treatment of fungicide. The results have clearly indicated that the leaf extracts of plants which are commonly available are antifungal in their property and helpful in reducing the rice blast disease in field and in increasing the yield. Though the plant derivatives are not as effective as fungicide in reducing the disease severity they were able to give a cost benefit ratio which is comparable to chemical fungicide. Plants like *Prosophis juliflora* and neem (*A. indica*) are commonly available in all villages and the marginal and submarginal can use them without much expenditure. These products not only reduce the input cost but are also unlikely to leave any harmful effect to the environment. Considering these facts, the farmers can use the pesticides of plant origin easily and relatively at a low cost for the management of rice blast disease of rice.

REFERENCES

Bisessar, S. 1965. A review of reaction of British Guyana rice varieties to blast disease. *Int. Rice Comm. Newsl.*, **14**: 31-33.

Cramer, H.H. 1967. Plant protection and World crop production. *Pflshultz. Nachr., Bayer,* **20**: 1-254.

Garcia, J., C. Morin, B. Nyasse, B.L. Sondergam, J.C. Tchouaukeu and E. Tsamo. 1991. Complete structure assignments of an ergosterol derivative from *Entaudrophragma utile. J. Natural Products,* **54**(1): 136-142.

Padmanabhan, S.Y. 1963. Estimating losses from rice blast in India. pp. 203-221. IRRI. The rice blast disease. Baltimore, John Hopkins Press, U.S.A.

Krishnamohan, G., P. Narayanasamy, K. Pillayarsamy and R. Jeyarajan. 1988. Efficacy of certain plant products on the rice blast pathogen *Pyricularia oryzae.* Natl. Seminar on Management of crop diseases with plant products/biological agent. Tamil Nadu Agrl. Univ. Madurai, Jan. 11-12 (1).

Lapis, D.D. and E. Dumancas. 1979. Survey of higher plants for fungicidal properties against *Pyricularia oryzae* Cav. *Philippine. Phytopath.,* **15**(1): 23-24.

Miah, M.A.T., H.D. Ahmed, N.R. Sharma, A. Ali, and S.A. Miah. 1990. Antifungal activity of some plant extracts. *Bangladesh J. Bot.,* **19**(1): 5-10.

Mishra, M., S.S. Malik, S.N. Tewari. 1992. Allelopathic effect of certain botanicals against the fungal pathogens of rice. *In:* Proc. First *National symposium* Allelopathy in agroecosystem (Agriculture and Forestry) Feb. 12-14, held at CSS, HAU, Hissar, India.

Mishra, M. and S.N. Tewari. 1990. Ethanolic extract toxicity of three botanicals against five fungal pathogens of rice. *Natl. Acad. Sci. Lr.,* **13**(11): 409-412.

Muthusamy, M., K. Pillayarsamy and V. Mariappan. 1988. Efficacy of plant products against blast and brown spot diseases of rice. Natl. Seminar on Management of crop diseases with plant products/-Biological agents. Madurai, India, Jan. 11-12 (21).

Rangaswami, G. and T.V. Subramanian. 1957. Estimation of losses caused by blast disease of rice. *Sci. Cul.,* **23**(4): 192-193.

Tewari, S.N. and Mandakini Nayak. 1991. Activity of four plant leaf extracts against three fungal pathogens of rice. *Trop. Agric.,* **68**(4): 373-375.

Tewari, S.N. and A. Premalatha Dath. 1984. Effect of leaf extract media of some plants on the growth of fungal pathogen of rice. *Indian Phytopath.,* **37**(3): 458-461.

Neem for the Management of Crop Diseases, 1995
Editor : V. Mariappan
Associated Publishing Co., New Delhi, India
pp. 11-17.

FIELD EVALUATION ON EFFICACY OF NEEM PRODUCTS ON MANAGEMENT OF FUNGAL DISEASES OF RICE

Yesuraja, I., Mariappan, V. and Sethuraman, K.

Department of Plant Pathology, Tamil Nadu Agricultural University
Coimbatore-641 003, Tamil Nadu

Rice crop is invariably affected by diseases caused by fungi and some of the diseases like brown spot (*Helminthosporium oryzae*), sheath rot (*Sarocladium oryzae*), sheath blight (*Rhizoctonia solani*) and grain discolouration (*Helminthosporium oryzae* and *Sarocladium oryzae*) are common in Tamil Nadu. These diseases cause damage to almost all varieties at all seasons resulting into considerable loss in yield. For control of these diseases, farmers of this area used to apply fungicides either individually or in combination with insecticides involving additional input cost. As a result the net gain realised by the farmers become reduced to a considerable extent besides causing pollution and ecological hazards. Neem and other plant products have been reported to be effective in controlling the rice diseases like blast (*P. oryzae*), sheath rot (*S. oryzae*) and sheath blight (*R. solani*) (Krishnamohan *et al.,* 1988; Viswanathan, 1989; Rajeswari, 1991). Hence the neem products like neem cake (NC), neem seed kernel extract (NSKE) and neem oil (NO) were compared with the recommended pesticides (insecticide and fungicide) simultaneously at different locations of rice growing areas in Tamil Nadu in a programme sponsored by Tamil Nadu Agricultural University - Asian Development Bank and International Rice Research Institute, Philippines. The results obtained in the programme as well as from those obtained from field trials conducted at other locations have been summarised below.

TREATMENTS

Nursery : In one set of plots in the nursery neem cake was applied as basal before last ploughing at 150 kg/ha and incorporated. Carbofuran was applied in another set of plots (at 1.4 kg/8 cents of nursery) 10 days after sowing. Carbofuran was applied mainly to control the insect vector (GLH) of rice tungrc virus disease. Control plots were not applied with any material. All plots were applied with recommended doses of manures and fertilizers.

Mainfield : Seedlings obtained from neem cake applied nursery plots were transplanted in separate plots for the treatments of neem oil (NO - 3%) and neem seed kernel extract (NSKE - 5%). Similarly seedlings obtained from carbofuran applied plots were transplanted in separate plots for the treatments with insecticides monocrotophos (0.1%) or phosphamidon (0.1%) or fungicide mancozeb (0.3%) depending upon the incidence of insect pest or disease. The spray preparations were applied at 500 l/ha using high volume sprayer. The spraying was given immediately after noticing the incidence of disease or insect pest exceeding the ETL. It was followed by the second spray and third spray at 15 days interval.

Preparation of NO and NSKE : Neem oil (3 l) was mixed with 100 litres of water containing liquid soap (at 0.1 per cent) and used for spraying. Neem seed kernel extract (5 kg) was powdered and soaked in 100 l of water for 8 hours. Then the content was filtered using a coarese cloth. The decanted liquid was added with liquid soap at 0.1% concentration and used for spraying.

RESULTS

The incidence of brown spot disease (*Helminthosporium oryzae*) alone was noticed at four locations (Coimbatore, Madurai, Puthuthamaraipatti, and Kondaiampatti) but the severity was relatively less at all places and differences between treatments were not significantly at most of the locations (Table 1). However the plots treated with neem products neem cake at nursery followed by neem oil or neem seed kernel extract spray recorded higher yield than the other treatments. The treatment neem cake followed by neem oil spray recorded maximum yield of 6030, 3800 and 3640 kg/ha at Madurai, Puthuthamaraipatti and Kondaiampatti respectively. The net gain achieved after meeting the expenses has been relatively higher in treatments involving neem preparations than the insecticidal or fungicidal treatments.

Table 1. Efficacy of neem products on brown spot disease of rice

Sl. No.	Location	Treatment	Brown spot (Mean grade)	Yield kg/ha	Yield value (Rs.)	Treatment cost (Rs.)	Net gain (Rs.)
1.	Coimbatore	NC + NSKE	1	3347[a]	10878	464	10414
		NC + NO	1	2974[c]	9665	389	9276
		Phosphamidan	1	3210[b]	10434	360	10074
		Control	3	2649[d]	8609	—	8609
		CD	NS				
2.	Madurai (ADT 36) (June '91-Sep. '91)	NC + NSKE	1.3	5110[b]	16608	1393	15215
		NC + NO	1.3	6030[a]	19598	1168	18430
		Monocrotophos/	1.2	4960[c]	16120	839	15281
		Mancozeb Control	1.6	2860[d]	9295	—	9295
		CD	NS	171			
3.	Puthuthamaraipatti (ADT 36) (July '91 - Oct. '91)	NC + NSKE	0.6	3400	11050	905	10145
		NC + NO	0.5	3800	12350	754	11956
		Monocrotophos/	0.4	3400	11050	606	10444
		Mancozeb Control	0.5	3300	10725	—	10725
		CD	NS	NS	—		

Sl. No.	Location	Treatment	Brown spot (Mean grade)	Yield kg/ha	Yield value (Rs.)	Treatment cost (Rs.)	Net gain (Rs.)
4.	Kondaiampatti	NC + NSKE	0.6	3000	9750	1344	8406
	(ADT 36)	NC + NO	0.6	3640	11830	1114	10716
	(July '91 -Aug. '91)	Monocrotophos +	0.5	2560	8320	829	7491
		Mancozeb Control	1.1	1640	5330	—	5330
		CD	0.1	60			

At Maruthuvakkudi incidence of brown spot (*H. oryzae*) and sheath blight (*Rhizoctonia solani*) were noticed while at Vinayagapuram, sheath rot disease (*Sarocladium oryzae*) alone was noticed (Table 2). The severity of brown spot disease was low and the differences between treatments have been not significant, while the differences between the treatments in respect of sheath rot and sheath blight were statistically significant.

Table 2. Effect of neem products on brown spot, sheath blight and sheath rot diseases of rice

Sl. No.	Location	Treatment	Brown spot (Mean grade)	Sheath blight (Mean grade	Sheath rot (PDI)	Yield kg/ha	Yield value (Rs.)	Treat-ment cost (Rs.)	Net gain (Rs.)
1.	Maruthuvakudi	NC + NSKE	1.0	3.0	—	4000	13000	464	12536
	(ADT 38)	NC + NO	1.3	1.9	—	4125	13406	389	13017
	(Sep. '91 - Feb. '92)	Monocrotophos/ Mancozeb	0.5	0.9	—	4300	13975	360	13615
		Control	1.5	4.9	—	3475	11294	—	11294
		CD	NS	0.4	—	187			
2.	Vinayagapuram	NC + NSKE	1.1	—	23.6	3452	11219	1344	9875
	(ADT 36)	NC + NO	1.0	—	36.4	3088	10036	119	8917
	(March '92 - June '92)	Monocrotophos/ Mancozeb	0.6	—	16.7	3448	11206	750	10456
		Control	1.6	—	35.8	1664	5408	—	5408
		CD	0.6	—	4.9	396	—		

At both locations plots treated with insecticide and fungicide recorded maximum yield and maximum net gain. This was followed by treatments involving neem products in recording increased yield and net gain.

At both locations plots treated with insecticide and fungicide recorded maximum yield and maximum net gain. This was followed by treatments involving neem products in recording increased yield and net gain.

Incidence of brown spot (*H. oryzae*) and grain discolouration (*H. oryzae* and

S. oryzae) were reported from the field trials conducted at Aduthurai (I and II season) (Table 3). The severity of these diseases were not significant different between treatments at both seasons (Table 3). However the yield recorded from plots treated with neem products or insecticide/fungicide were significantly higher than those obtained from control. The net gain obtained from treatments of insecticide/fungicide was the highest followed by the treatments involving neem products. They were considerably higher than the net gain obtained from control.

Table 3. Efficacy of neem products on brown spot and grain discolouration diseases of rice

Sl. No.	Location	Treatments	Mean Grade		Yield (kg/ha)	Yield value (Rs.)	Treat-ment cost (Rs.)	Net gain (Rs.)
			Brown spot	Grain discolou-ration				
1.	Aduthurai I season (ADT 38) (Sep. '91 - Feb. '92)	NC + NSKE	1.5	2.0	5000[a]	16250	464	15786
		NC + NO	1.5	1.9	5125[a]	16656	389	16267
		Carbofuran + Monocrotophos/ Mancozeb	2.7	2.1	5300[a]	17225	360	17189
		Control	2.9	2.3	4305[a]	13991	—	13991
		CD	NS	NS	597			
2.	Aduthurai II season (ADT 39) Nov. '91 - Feb. '92)	NC + NSKE	1.7	1.2	4630[a]	15048	464	14584
		NC + NO	1.7	1.5	4839[a]	15727	389	15338
		Carbofuran + Manocrotophos/ Mancozeb	1.8	1.6	4908[a]	15951	360	15591
		Control	1.6	1.3	3177[b]	10325	—	10325
		CD	N.S	321				

(Final report of the IRRI - ABD - TNAU - Collaborative project on Botanical Pest Control, 1992).

At Pasur the rice crop was affected by sheath blight disease and the severity of the disease was reduced considerably in plots treated with neem and chemical spray application when compared to control. The yield obtained from plots treated with insecticide has been found to be maximum (7667 kg/ha). It was followed by neem cake and neem seed kernel extract treatment which recorded an yield of 5684 kg/ha. However both of them were on par (Table 4).

Table 4. Efficacy of neem products on sheath blight disease of rice at Pasur

Sl. No.	Treatments	Sheath blight (PDI)	Yield (kg/ha)	Yield value (Rs.)	Treatment cost (Rs.)	Net gain (Rs.)
1.	NC + NSKE	25.18[a]	5684[a]	18473	18009	
2.	NC + NO	26.18[a]	5476[b]	17797	389	17408
3.	Carbofuran + Phosphamidon	30.68[b]	7667[a]	24918	360	24883
4.	Control	33.16[b]	4494[c]	14606	—	14606

Variety : ADT 36 Season : July '91 - Oct. '91

Integrated Pest Management trial on rice (IPM)

In the field trials conducted for the comparison of IPM practice (including NSKE as an alternative to pesticide) and IPM practice (including chemical pesticides) with farmers practice, incidence of rice blast (*Pyricularia oryzae*), brown spot (*H. oryzae*) and sheath rot (*S. oryzae*) diseases was considerably reduced in the IPM plots than that of the plots receiving plant protection measures as followed by farmers (Table 5). At all the five locations of Coimbatore, Gobi, Vaidyanathampatti, Rajendrum and Aduthurai the yield obtained per hectare was relatively more in plots receiving IPM practice than that of the plots receiving the treatments of as per farmers practice. The yield obtained from plots receiving insecticides and neem seed kernel extract as a component of IPM has been almost similar.

Table 5. IPM practices for the control of rice disease

Sl. No.	Treatments	Mean incidence (PDI)			% Increase or decrease farmers' method	Yield (kg/ha)				
		Blast	Brown spot	Sheath rot		Coimbatore	Aduthurai	Gobi	Vaidyanathampatti	Rajendrum
1.	Farmers practice	9.3^{d}	18.6^{b}	13.9^{b}	—	4298	7092	4446	5010	7316
2.	Complete IPM with insecticides	8.7^{c}	17.5^{a}	12.8^{b}	14.70	5187	7971	5000	7428	9064
3.	Complete IPM with NSKE	7.8^{b}	16.8^{a}	10.7^{a}	11.65	4964	7820	4964	7242	8730

DISCUSSION

Rice crop is affected by many diseases. In U.S.A., the over all loss from diseases of rice has been estimated at six per cent of the crop yield (Cramer, 1967). In India the loss caused by a single disease 'blast' alone has been reported to be as high as 70 per cent when there was severe incidence (Rangaswami and Subramanian, 1957). As the crop is affected by diseases at different stages of its growth, the loss in yield due to diseases alone is quite considerable. The farmers invariably use fungicide specific for each disease/pathogen and thus there are occassions where many types of fungicides are applied to a single crop involving high input cost, leading to ecological hazards etc.

The efficacy of plant products on several plant pathogens have been reported in recent years. These substances have been reported to contain the diseases caused by them on plants either by directly acting on the causal pathogens or by inducing resistance in host plants resulting in reduction in disease development (Bhowmick and Vardhan, 1981; Devi *et al.,* 1982; Chaturvedi *et al.,* 1987; Datar, 1988; Doubrava *et al.,* 1988).

Manian *et al.* (1988) reported that cold water extracts of *Prosophis juliflora,*

Lantana camara and *Azadirachata indica* were inhibitory to the germination of the spores of rice brown spot pathogen (*Helminthosporium oryzae*). Pan (1990) observed in the slide germination and growth inhibition tests with *Drechslera oryzae* the addition of *Azadirachta indica* or *Pongamia glabra* oils to captan increased the fungicidal activity synergistically. Neem (*A. indica*) extract has been also reported to inhibit the growth of the rice sheath rot pathogen (*Sarocladium oryzae*) (Easwaramoorthy, 1989). Neem oil (at 2.5 and 5.0 per cent concentrations) has been reported to be inhibitory to the growth of *Drechslera rostratum, Macrophomina phaseoli* and other plant pathogens (Vir and Sharma, 1985). The severity of brown spot disease of rice has been reported to be reduced in plots sprayed with neem oil and neem seed kernel extract (Thangamani Narayanasamy and Narayanasamy, 1988). Neem oil (one per cent) and neem seed kernel extract (two per cent) were reported to be effective in reducing rice sheath rot (*S. oryzae*) disease (Mariappan *et al.,* 1988; Narasimmaraj, 1991). Intensity of sheath blight disease (*Rhizoctonia solani*) has been reported to be reduced in plots amended with non eible seed cakes including neem (Rajan, 1980). The field trials conducted at farmers field in many locations have indicated clearly the usefulness of neem preparations for the management of rice diseases. Neem products are commonly available to the native farmers and they can use these preparations relatively at a cheaper cost and can reduce the input expenditure easily. The cost benefit ratio obtained in many trials have clearly indicated that the neem preparations are comparable with the chemical pesticides. Neem seed kernel extract preparations have been found to be suitable substitute for the pesticides in the Integrated Pest management practice.

REFERENCES

Bhowmick, B.N. and N. Vardhan. 1981. Antifungal activity of some leaf extracts of medicinal plants on *Curvularia lunata. Indian Phytopath.,* **34**: 385-386.

Chaturvedi, R., D. Anupam, C. Dikshit and S.N. Dixit. 1987. *Adenocalyma allicea* a new source of natural fungitoxicant. *Trop. Agric.,* **64**(4): 318-382.

Cramer, H.H. 1967. Plant protection and world crop production. *Pflshultz. Nachr., Bayer,* **20**: 1-254.

Datar, V.V. 1988. Antifungal activity of plant extracts against *Fusarium* sp. causing rhizome rot of ginger. *Nat. Sem. on Management of crop diseases with plant products/biological agents.* Agricultural College and Res., Instt., Madurai, pp. 28-29. (Abst.)

Devi, S.B., A. Naseema and M.C. Nair. 1982. *In vitro* effect of lemon grass oil on the mycelial growth of *Rhizoctonia solani. Indian Phytopath.,* **35**: 714-716.

Doubrava, N.S., A. Ralph and Dean H. Kuc. 1988. Induction of resistance to anthracnose caused by *Colletotrichum lagenarium* in cucumber by oxalate and extracts from spinach and rhubarb leaves. *Physiol. Mol. Plant Pathol.,* **33**(1): 69-79.

Easwaramurthy, S., M. Muthusamy, S. Muthusamy and V. Mariappan. 1989. Inhibitory effect of neem, *Acacia, Prosophis* and *Ipomoea extracts* on *Sarocladium oryzae* and *Fusarium oxysporum* f. sp. *cepae. Neem Newsl.,* **6**: 4-5.

Krishnamohan, G., P. Narayanasamy, K. Pillayarsamy and R. Jeyarajan. 1988. Efficacy of certain plant products of certain plant products on the rice blast Pathogen *Pyricularia oryzae. Nat. Sem. on Management of crop diseases with plant products/biological agents.* Agricultural college and Res. Inst., Madurai, pp. 1. (Abst.).

Manian, S., P. Lakshmanan and K. Raju. 1988. Effect of six plant extracts on spore germination of *Helminthosporium oryzae* Breda De Mann. Nat. Sem. on Management of crop diseases with plant products/biological agents. Agricultural College & Res. Inst., Madurai, pp. 4. (Abst.).

Mariappan, V., V. Narasimhan and S. Jayaraj. 1988. Effect of neem on rice diseases. Proc. of final workshop of IRRI-ADB-EWC project on Bothanical pest control in Rice based cropping systems held in Dec. 12-18, 1988. *Int. Rice Res. Inst.,* Philippines. pp. 8.

Narasimmaraj, V. 1991. Studies on rice sheath rot disease caused by *Sarocladium oryzae* Sawada. M.Sc.(Ag.) Thesis, Tamil Nadu Agric. Univ., Coimbatore, India, pp. 138.

Pan, S. 1990. Additive effects of plant oils and fungicides. *J. Mycopathol. Res.,* **28**(1): 57-61.

Rajan, K.M. 1980. Soil amendments in plant disease control. *Int. Rice Res. Newsl.,* **5**(4): 15.

Rajeswari, E. 1991. Effect of plant derivatives on rice blast pathogen (*Pyricularia oryzae* Cav.) M.Sc.(Ag.) Thesis, Tamil Nadu Agric. Univ., Coimbatore, India. pp. 134.

Rangàswami, G. and T.V. Subramanian. 1957. Estimation of losses caused by blast disease of rice. *Sci. Cul.,* **23**(4): 192-193.

Thangamani Narayanasamy and P. Narayanasamy. 1988. Role of neem products in the control of fungal diseases of rice. Nat. Sem. on Management of crop diseases with plant products/biological agents. Agricultural College and Res. Inst., Madurai, pp. 6-7. (Abst.).

Vir, D. and R.K. Sharma. 1985. Efficacy of fungicides, XXXII. Evaluation of neem oil for control of plant pathogens. *Asian Farm Che.,* **1**(7/8): 23-24.

Viswanathan, R. 1989. Management of rice fungal diseases with mancozeb and tricyclazole. M.Sc.(Ag.) thesis, Tamil Nadu Agric. Univ., Coimbatore, India, pp. 85.

Neem for the Management of Crop Diseases, 1995
Editor : V. Mariappan
Associated Publishing Co., New Delhi, India
pp. 19-26.

EFFICACY OF DIFFERENT PLANT PRODUCTS ON BROWN SPOT OF RICE CAUSED BY *DRECHSLERA ORYZAE*

Devanathan, M. Sree Ramkumar, R. and Narayanasamy, R.

Department of Plant Pathology, Faculty of Agriculture, Annamalai University
Annamalai Nagar, Tamil Nadu

Brown spot disease of rice caused by *Drechslera oryzae* (Breda de Haan) Subram and Jain (Syn. *Helminthosporium oryzae*) is an important disease causing yield loss of 19.2-41.2 per cent in cases of severe infection (Vidhyasekaran and Ramadoss, 1973). One of serious outbreaks was in Bengal in 1942 when loss in yield upto 90 per cent was recorded (Ghose *et al.,* 1960).

Increased and indiscriminate use of various plant protection chemicals, including fungicides, has been causing great environmental hazards in recent years. Therefore it is preferable to use alternative control measures when ever possible (Tarr, 1972).

Green plants, particularly tropical flora act as a reservoir of various biologically active substances. Surveys and screening of higher plants for antimicrobial properties carried out in different countries have demonstrated the occurrence of antifungal substances in higher plants (Zineb *et al.*, 1959; Nene *et al.*, 1968; Misra and Dixit, 1976; Tripathi *et al.*, 1978 and Bhaskaran, 1990).

Fungitoxicants from higher plants possibly may have a better future over synthetic fungicides due to their largely non-phytotoxic, systemic and easily biodegradable nature (Fawcett and Spencer, 1970 and Beye, 1978).

The potential of fungitoxicity of several plant species have been well demonstrated in recent years and some of these are presented below.

Plant extracts were reported to be inhibitory to the spore germination mycelial growth and biomass production.

Sreeramakumar (1990) reported that among the 23 plant extracts tested at 10 per cent concentration, *Lawsonia inermis* showed the maximum inhibition with only 9.33 per cent spore germination, while the least inhibition was observed in *Thevetia peruviana* with 81.00 per cent spore germination (Table 1).

Table 1. Effect of cold water extracts of different plant species at ten per cent concentration on the spore germination of *Drechslera oryzae* (Sreeramakumar, 1990)

	Treatment	Percentage of spore germination 10 % concentration		
		Mean	Comparison	Per cent increase (+)/ decrease (–) over control
T_1	*Arbus piecatorius*	38.00	j	– 58.99
T_2	*Adhatoda vasica*	18.67	p	– 79.85
T_3	*Allium cepa*	35.67	k	– 61.51
T_4	*Azadirachta indica*	10.67	s	– 88.49
T_5	*Bassia latifolia*	30.00	l	– 67.63
T_6	*Bougainvillaea spectabilis*	45.67	i	– 50.72
T_7	*Caesalpinia pulcherrima*	50.00	h	– 46.05
T_8	*Calotropis gigantea*	13.67	r	– 85.25
T_9	*Carica papaya*	50.67	h	– 45.32
T_{10}	*Cassia fistula*	21.00	n	– 77.34
T_{11}	*Cocos nucifera*	19.00	pq	– 79.50
T_{12}	*Croton sparsiflorus*	59.33	g	– 35.98
T_{13}	*Eucalyptus globulus*	12.00	s	– 87.05
T_{14}	*Ficus retusa*	70.00	f	– 24.46
T_{15}	*Ipomoea carnea*	20.35	nop	– 78.06
T_{16}	*Lawsonia inermis*	9.33	t	– 89.93
T_{17}	*Morinda tinctoria*	25.67	m	– 72.30
T_{18}	*Nerium indicum*	72.33	e	– 21.95
T_{19}	*Ocimum sanctum*	31.00	l	– 66.55
T_{20}	*Pedilanthes tithymaloides*	79.00	c	– 14.75
T_{21}	*Pongamia glabra*	21.00	no	– 77.34
T_{22}	*Solanum torvum*	77.00	d	– 16.91
T_{23}	*Thevetia peruviana*	81.00	b	– 12.59
T_{24}	Bavistin 50 WP (0.25%)	9.00	t	– 90.29
T_{25}	Control	92.67	a	–
	Mean			
	SE		0.7017	
	CD		1.3753	

* Mean of three replications.
Note : In column, means followed by the same letters are not significantly different.

Azadirachta indica, Eucalyptus globulus, Calotropis gigantea, Adhatoda vasica and *Cocos nucifera* were the other plant extracts which showed spore germination percentage below 20.00.

Sreeramakumar (1990) also reported that among the plnat extracts tested *Lawsonia inermis* 10 per cent concentration recorded the maximum inhibition of growth of the pathogen with the mean diameter of 16.00 mm, which was 82.22 per cent decrease over control. *Adhatoda vasica* registered the minimum inhibition with the mean diameter of 81.67 mm.

Azadiracta indica, Eucalyptus globulus, Calotropis gigantea, Cassia fistula, Pongamia glabra, Morinda tinctoria, Cocos nucifera and *Ipomoea carnea* recorded 20.00, 21.67, 26.67, 36.67, 39.33, 43.33, 47.67 and 77.67 mm diameter of growth respectively (Table 2). Bavistin at 0.25 per cent concentration showed maximum inhibition of growth with the mean diameter of 8.00 mm, which was 91.11 per cent less over control. In control the radial growth was found to be 90.00 mm in diameter.

Table 2. Effect of cold water extracts of ten selected plant species at ten per cent concentration on the radial growth of *D. oryzae* (Sreeramakumar, 1990)

Treatment		Diameter of growth mm 10 % concentration		
		Mean	Comparison	Per cent increase (+)/ decrease (–) over control
T_1	*Adhatoda vasica*	81.67	b	– 9.26
T_2	*Azadirachta indica*	20.00	i	– 77.78
T_3	*Calotropis gigantea*	26.67	h	– 70.37
T_4	*Cassia fistula*	36.67	g	– 59.26
T_5	*Cocos nucifera*	47.67	d	– 47.03
T_6	*Eucalyptus globulus*	21.67	i	– 75.92
T_7	*Ipomoea carnea*	77.67	c	– 13.70
T_8	*Lawsonia inermis*	16.00	i	– 82.22
T_9	*Morinda tinctoria*	43.33	e	– 51.86
T_{10}	*Pongamia glabra*	39.33	f	– 56.30
T_{11}	Bavistin 50 WP (0.25%)	8.00	k	– 91.11
T_{12}	Control	90.00	a	–
	Mean	42.39		
	SE		1.4845	
	CD		2.9626	

* Mean of three replications.
Note : In column, mean followed by the same letter are not significantly different

The results from Table 3 showed that *Lawsonia inermis* at 10 per concentration registered maximum inhibition in growth with the mean mycelial dry weight of 30 mg per 50 ml of broth which was on par with bavistin.

Table 3. Effect of cold water extracts of ten selected plant species at ten per cent concentration on the mycelial dry weight of *D. oryzae* (Sreeramakumar, 1990)

Treatment		Mycelial dry weight mg/50 ml broth 10 % concentration		
		Mean	Comparison	Per cent increase (+)/ decrease (–) over control
T_1	*Adhatoda vasica*	151	d	– 39.60
T_2	*Azadirachta indica*	64	b	– 74.40
T_3	*Calotropis gigantea*	85	g	– 66.00
T_4	*Cassia fistula*	185	c	– 26.00
T_5	*Cocos nucifera*	125	t	– 50.00
T_6	*Eucalyptus globulus*	57	i	– 77.20
T_7	*Ipomoea carnea*	200	b	– 20.00
T_8	*Lawsonia inermis*	30	j	– 88.00
T_9	*Morinda tinctoria*	190	c	– 24.00
T_{10}	*Pongamia glabra*	141	e	– 43.00

Treatment		Mycelial dry weight mg/50 ml broth 10 % concentration		
		Mean	Comparison	Per cent increase (+)/ decrease (–) over control
T_{11}	Bavistin 50 WP (0.25%)	30	j	– 88.00
T_{12}	Control	250	a	–
	Mean			
	SE		2.5386	
	CD		5.0671	

* Mean of three replications.

Note : In column, mean followed by the same letter are not significantly different

Eucalyptus globulus, Azadirachta indica, Calotropis gigantea, Cassia fistula, Morinda tinctoria and *Ipomoea carnea* recorded 57, 64, 85, 125, 141, 151, 185, 190 and 200 mg of mycelial dry weight per 50 ml of broth respectively.

In the case of seed germination (Table 4), *Adhatoda vasica, Cocos nucifera* and *Lawsonia inermis* recorded the maximum percentage of seed germination i.e., 95 which was on par with Bavistin.

Table 4. Effect of cold water extracts of different plant species on the seed germination of rice cultivar IR 20 - Rolled towel method (Sreeramakumar, 1990)

Treatment		Percentage of seed germination 10 % concentration	
		Mean	Per cent increase (+)/decrease (–) over control
T_1	*Arbus precatorius*	870	– 1.14
T_2	*Adhatoda vasica*	95	+ 7.95
T_3	*Allium cepa*	87	– 1.14
T_4	*Azadirachta indica*	93	5.68
T_5	*Bassia latifolia*	88	0.00
T_6	*Bougainvillaea spectabilis*	88	0.00
T_7	*Caesalpinia pulcherrima*	80	– 9.09
T_8	*Calotropis gigantea*	91	– 3.41
T_9	*Carica papaya*	86	-h2.27
T_{10}	*Cassia fistula*	90	2.27
T_{11}	*Cocos nucifera*	95	7.95
T_{12}	*Croton sparsiflorus*	85	– 3.40
T_{13}	*Eucalyptus globulus*	92	4.55
T_{14}	*Ficus retusa*	82	– 6.82
T_{15}	*Ipomoea carnea*	90	2.27
T_{16}	*Lawsonia inermis*	95	7.95
T_{17}	*Morinda tinctoria*	90	2.27
T_{18}	*Nerium indicum*	82	– 6.82
T_{19}	*Ocimum sanctum*	89	1.17
T_{20}	*Pedilanthes tithymaloides*	85	– 3.40
T_{21}	*Pongamia glabra*	91	3.41
T_{22}	*Solanum torvum*	81	7.95
T_{23}	*Thevetia peruviana*	82	– 6.82

	Treatment	Percentage of seed germination 10 % concentration	
		Mean	Per cent increase (+)/decrease (–) over control
T_{24}	Bavistin 50 WP (0.25%)	95	7.95
T_{25}	Control	88	—
	Mean		
	SE		0.8266
	CD		1.6425

* Mean of three replications.

Caesalpinia pulcherrima recorded the lowest percentage with only 80.

In field experiment with IR 50 (Table 5) showed that all the treatments were found to reduce the percentage of disease incidence conspicuously when compared to control. Hinosan registered the maximum reduction of 37.22 per cent over control. The plant extracts, *Lawsonia inermis, Azadirachta indica, Eucalyptus globulus* and *Calotropis gigantea* recorded 31.17, 26.08, 22.31 and 18.39 per cent reduction over control respectively.

Table 5. Effect of cold water extracts of four selected plant species on the percentage of brown spot disease incidence and yield of rice cultivar IR 50 (Sreeramakumar, 1990)

	Treatments	Percentage of disease incidence			Grain yield tonnes/ha		
		Mean	Comparison	Per cent increase (+)/decrease over control	Mean	Comparison	Per cent increase (+)/decrease over control
T_1	*Azadirachta indica* (10%)	30.19 (33.33)	d	– 26.08	6.77	abc	27.74
T_2	*Calotropis gigantea* (10%)	33.33 (35.26)	b	– 18.39	6.33	cd	19.43
T_3	*Eucalyptus globulus* (10%)	31.73 (34.28)	c	– 22.31	6.68	abcd	26.04
T_4	*Lawsonia inermis* (10%)	28.03 (31.97)	e	– 31.37	7.05	ab	33.02
T_5	Hinosan 50 EC (0.1%)	25.64 (30.42)	f	– 37.33	7.17	a	35.28
T_6	Control	40.84 (39.72)	a	–	5.30	e	–
	SE		0.2691		0.2682		
	CD		0.5995		0.5976		

* Mean of three replications.

Note : Means followed by the same letter are not significantly different.
Figures in parentheses are angular transformed values.

The grain yield also very high in *Lawsonia inermis, Azadiracta indica, Eucalyptus globulus* and *Calotropis gigantea* treated plots. They registered 7.05, 6.77, 6.68

and 6.33 tonnes/ha respectively. Hinosan recorded the maximum grain yield of 7.17 tonnes/ha.

There was no significant difference between the disease incidence and yields due to Hinosan, *Lawsonia inermis, Eucalyptus globulus* and *Azadirachta indica* application.

Out of the twenty three plant species, ten effective ones viz., *Adhatoda vasica, Azadirachta indica, Calotropis gigantea, Cassia fistula, Eucalyptus globulus, Ipomoea carnea, Lawsonia inermis, Morinda tinctoria* and *Pongamia glabra* were retained for further studies.

The cold water extract of *Lawsonia inermis* at 10 per cent concentration was on par with Bavistin. Few reports are available on the antifungal efficacy of *Lawsonia inermis* (Natarajan and Lalithakumari, 1987); *Azadirachta indica* (Manian *et al.*, 1988) and *Bougainvillaea* sp. (Varaprasad, 1988) on the conidial germination.

The minimum mean diameter of radial growth was recorded at 10 per cent concentration. The same trend was noted in the case of reduction in mycelial dry weight also where the minimum mycelial dry weight was registered with 10 per cent concentration.

The maximum inhibition of radial growth among plant extracts was recorded with *Lawsonia inermis* followed by *Azadirachta indica, Eucalyptus globulus, Calotropis gigantea, Cassia fistula, Pongamia glabra, Morinda tinctoria, Ipomoea carnea* and *Adhatoda vasica.*

In the case of mycelial dry weight, the effective plant extracts in the order of merit were, *Lawsonia inermis, Azadirachta indica, Eucalyptus globulus, Calotropis gigantea, Cocos nucifera, Pongamia glabra, Adhatoda vasica, Morinda tinctoria, Cassia fistula* and *Ipomoea carnea.* This fungitoxic effect on radial growth and mycelial dry weight was reported by Unnikrishna Pillai (1988) and Varaprasad (1988) on the growth of *Drechslera oryzae.*

Cold water extracts of different plant species of 10 per cent concentration as well as the fungicide Bavistin 50 wp at 0.25 per cent concentration exhibited different effects on the seed germination. The effect of seed treatment using plant extracts/products was studied by numerous researchers (Jeeva, 1989; Shetty *et al.*, 1989). The enhancement of seed germination by the plant extracts might be due to several factors such as the fungitoxic action leading to the killing of the pathogens present both internally and externally in the seeds, the presence of auxins in the plant extracts.

The foliar application of the cold water extracts of all the four plant species at 10 per cent concentration and fungicides Bavistin and Hinosan at 0.25 and 0.1 per cent concentrations respectively checked the brown spot disease incidence significantly in IR 50 variety (Asthana *et al.,* 1986).

In the field trial with IR 50, the best among the plant extract was *Lawsonia inermis* followed by *Azadirachta indica, Eucalyptus globulus* and *Calotropis gigantea.* All were superior to control.

The above over all superiority of *Lawsonia inermis* extract in disease control might be due to the presence of the antifungal factor, namely 'Lawsone' (2-hydroxy-1, 4-naphtoquinone) in the leaves of this plant (Tripathi *et al.,* 1978).

Thangamani Narayanasamy and Narayanasamy (1988) observed that neem oil and neem seed kernel extract treatment recorded significantly higher yield and decreased the disease incidence than untreated control in rice. This finding lends support to the present work.

REFERENCES

Asthana, A., N.N. Tripathi, and S.N. Dixit, 1986. Fungitoxic and phototoxic studies with essential oil of Ocimum adacendens. *J. Phytopath.* **117**(2): 152-159.

Beye, I. 1978. Insecticides from the vegetable kingdom. *Plant Res. Dev.,* **7**: 13-31.

Bhaskaran, R. 1990. Biological control of Thanjavur wilt disease of coconut. Abstracts of papers of *National symposium on Biocontrol of root diseases,* Annamalai Univ. Annamalai Nagar, Jan. 2-4, pp. 7-8.

Fawcett, C.H. and D.M. Spencer, 1970. Plant chemotherapy with natural products. *Ann. Rev. Phytopathol.,* **8**: 403-418.

Ghose, R.I.M., M.B. Ghatge, and V. Subrahmanyan. 1960. Rice in India (Re. edn.), ICAR, New Delhi, pp. 474.

Jeeva, M.L. 1989. Studies on the effect of certain plant products on the growth, sporulation and spore germination of *Fusarium oxysporum* fsp. *lycopersici* (Scc.) Snyder and Hanen and their application with *S. oryzae* inoculation on seed treatment, sheath rot disease incidence and yield parameters of rice (*Oryza sativa* L.) M.Sc.(Ag.) Thesis, Annamalai Univ., India.

Manian, S., P. Lakshmanan, and K. Raju, 1988. Effect of six plant extracts on spore germination of *Helminthosporium oryzae* Breda de Hann. Abstracts of papers of *National seminar on management of crop disease with plant products/biological agents.* Tamil Nadu Agr. Univ., Madurai, Jan. 11-12. p. 4.

Misra, S.B. and S.N. Dixit, 1976. Fungicidal spectrum of the leaf extract of *Allium sativum. Indian Phytopath.,* **29**: 448-449.

Natarajan, M.R. and D. Lalitha Kumari, 1987. Leaf extracts of *Lawsonia inermis* as antifungal agent. *Curr. Sci.,* **56**: 101-102.

Nene, Y.L., P.N. Thaplial and S. Krishna Kumar, 1968. Screening of some plant extracts for antifungal properties. Lab dev. Jour. *Food Sci. and Tech.,* **68**(4): 221-228.

Shetty, S.A., H.S. Prakash, and H.S. Shetty, 1989. Efficacy of certain plant extracts against seed-borne infection of *Trichoniella padwickii* in paddy (*Oryza sativa*). *Can. J. Bot.,* **67**: 1956-1958.

Sreeramakumar, R. 1990. Efficacy of different plant products on brown spot of rice caused by *Drechslera oryzae* (Breda de Haan) Subram. and Jain. M.Sc.(Ag.). Thesis, Faculty of Agriculture, Annamalai University, India.

Tarr, S.A.J. 1972. *Principles of Plant Pathology.* Mc Millan Press, London and Basing stoke.

Thangamani Narayanasamy and P. Narayanasamy, 1988. Role of neem products in the control of fungal disease of rice. Abstracts of papers of *National seminar on management of crop diseases with plant products/biological agents.* Tamil Nadu Agri. Univ., Madurai, Jan. 11-12, p. 6.

Tripathi, R.D., H.S. Srivastava and S.N. Dixit, 1978. Lawsone, an antifugal antibiotic from leaves of *Lawsonia inermis;* Its action on pathogen and host plant, *Indian J. Mycol. and Pl. Pathol.,* 81: 89-90.

Unnikrishna Pillai, K.P. 1988. Studies on the effect of plant extracts/products and antagonistic bacterium against rice brown spot pathogen *Drechslera oryzae* (Breda de Haan) Subram and Jain. M.Sc.(Ag.) Thesis, Tamil Nadu Agri. Univ., India.

Varaprasad, M. 1988. Studies on the effect of certain plant production on seed and soil-borne pathogens (*Helminthosporium oryzae*) Breda de Hann, *Rhizoctonia solani* Kuhn. (*Thanatephorus cucumeris* (Frank) Donk.) *Aspergillus niger* Van Tiegh., *Aspergillus flavus* Link), M.Sc.(Ag.) Thesis, Annamalai Univ., India.

Vidhyasekaran, P. and N. Ramadoss, 1973. Quantitative and qualitative losses in paddy due to Helminthosporiose epidemic. *Indian Phytopath.,* 26: 479-484.

Zineb, F.I., Tobshy and J.B. Sinclaria, 1959. Inhibition *in vitro Phomopsis citri* by extracts from orange. *Plant Dis. Rep.,* 48: 925.

Neem for the Management of Crop Diseases, 1995
Editor : V. Mariappan
Associated Publishing Co., New Delhi, India
pp. 27-30.

NEEM IN THE CONTROL OF RICE BACTERIAL LEAF BLIGHT DISEASE

Eswaramurthy, S., Mariappan, V. and Alagianagalingam, M.N.

Department of Plant Pathology, Agricultural College and Research Institute
Tamil Nadu Agricultural University, Madurai-625 104, Tamil Nadu

INTRODUCTION

Neem (*Azadirachta indica* A. Juss) contains various antipathogenic chemicals and is of immense use in the control of various fungal, bacterial, and virus diseases of plants. Application of plant products and antagonists has a significant role in controlling diseases and in minimizing the environmental hazards due to use of certain chemicals. The role of Neem products in the control of rice bacterial blight disease is discussed.

Effect of green manures and neem cake

Incorporation of leaves of *Azadirachta indica,* L. Juss, *Prosophis juliflora* Dc, *Ipomea cornea* Jacq was highly effective in controlling Rice bacterial blight (BB) inicited by *Xanthomonas campestris* Pv *oryzae* (Table 1). These leaves are having bactericidal properties and through altering the host physiology they could have reduced bacterial blight. Application of bulky organic manure has profound influence on disease intensity because of their effect on soil microbial activities. Rajan and Reddy (1986) detected the influence of neem cake, castor cake, green leaf, farm yard manure on rice sheath blight. Santhi and Palaniappan (1989) reported that the application of green manures such as *Azadirachta indica* L. Juss leaves, *Pongamia glabra* Vent leaves increased rice plant vigour, yield and application of

Table 1. Effect of Green Manures and Neem Cake on Rice Bacterial Blight

Sl. No.	Treatment	Mean lesion length in cms.*
1.	*Azadirachta indica* leaves	2.8
2.	*Prosphis julifera* leaves	3.1
3.	*Ipomea cornea* leaves	5.3
4.	Neem cake soil application	2.6
5.	Control	12.4

* Mean of 4 replications
C.D. (P = 0.05) – 0.16

neem products inhibited nitrifying organisms. Thyagarajan *et al.* (1986) found that Diammonium phosphate at 44 Kgs/ha with green manures at 10T/ha recorded lower incidence of BB than with DAP alone.

The application of green manures increases the Soil microbial activities, Soil microbial population, activities of Soil enzymes such as dehydrogenases and thereby enhances the rice yield. Gopalsamy and Vidhyasekaran (1987) also have stated this fact. Temburnikar (1989) detected the reduction in bacterial blight due to green manuring.

Effect of neem coated urea, coal tar coated urea

Neem coated urea, coal tar coated urea sharply reduced baterial blight by recording mean lesion length of 3.1 cms, 8.3 cms as against 12.5 cms in prilled urea (Table 2) Through the bactericial effect, altered host physiology and through the slow release of nitrogen the neem incorporated urea could have brought about reduction in disease. Santhi and Palaniappan (1989) reported the efficacy of neem coated urea in increasing the vigour of rice plant and yield of rice. Fawcett and Spencer (1966) reported the antifungal and antibacterial effects of plant extracts and plant exudates.

Table 2. Effect of Neem Coated Urea, Coal Tar Coated Urea on Rice Bacterial Blight

Sl. No.	Treatment	Mean BB lesion length in cms.*
1.	Prilled urea	12.5
2.	Coal Tar Coated urea	8.3
3.	Neem coated urea	3.1
4.	Control	10.0
	* Mean of 5 replications C.D. (P = 0.05) –	0.72

Neem seed kernel extract, neem cake effect

Azadirachta indica A. Juss seed kernel extract 2 percent, cake extract 5 percent, 10 percent bulb extract of *Allium sativum* Linn, 10 percent leaf extract of *Prosophis Juliflora* Dc, *Acacia arabica* willd, *Ipomea cornea* Jacq inhibited the bacterial blight pathogen. Eswaramurthy (1979) reported that the root exudates of *Allium sativum* Linn and *Sorghum vulgare* pers strongly inhibited *Pseudomonas solanacearum. Punica granatum* leaf extracts exhibited antibacterial activity (Rao and George, 1949). Under field conditions spraying of 10 percent leaf extracts of *prosophis juliflora, Acacia arabica* Neem seed kernel extract 2 percent, *Bacillus Subtilis* showed mean bacterial blight intensity grade of 2.38, 2.88, 3.0, 2.0 only as against 5.75 in control (Table 3). Spraying of water extracts of these plant products, at boot leaf stage at forthightly interval effectively checked the bacterial blight. Baker and Cook

(1974) had well emphasized the significance of biological control. Neem products occupy a significant place in such an aspect in the control of various diseases. Rangaswami (1975) stated sprinkling neem cake suspension (1 Kg in 20 litres of water) 20 times a year resulted in some control of citrus canker on acid lime.

Table 3. Control of Rice Bacterial blight using Neem products and antagonist

Sl. No.	Treatment	Mean bacterial blight intensity grade			Mean yield in Kgs/ha		
		Rabi	Rabi	Mean	Rabi	Rabi	Mean
1.	Neem Cake extract 5 percent	1.00	5.00	3.0	2650	3207	2928.0
2.	Neem seed kernel extract	1.00	5.00	3.0	2950	3120	3035.0
3.	*Prosophis juliflora* leaf extract 10 percent	1.75	4.00	2.88	2650	3315	2982.5
4.	*Acacia arabica* extract 10 percent	1.75	4.00	2.88	2850	2395	2622.5
5.	Plantomycin 500 ppm	1.00	4.50	2.75	2850	3055	2952.5
6.	Plantomycin 1000 ppm	1.00	3.75	2.38	3100	2960	3030.0
7.	*Bacillus subtilis*	1.00	3.00	2.0	3000	2385,	2692.5
8.	Control	3.50	8.00	5.75	2300	1213	1756.5
	C.D. (P = 0.05)	0.87	1.68				

CONCLUSION

Application of neem products brought about variation in rice host physiology. Through the bactericidal effect, through altered host physiology through effect on other soil microbes and through the phytotonic effect on host plant etc. neem products significantly reduce the bacterial diseases and play a salient role in reducing the environmental hazards due to use of certain chemical. Neem products use is to be encouraged in order to overcome the pollution hazards and for the prevention of resistant strains of pathogens.

REFERENCES

Baker, K.F. and R.J. Cook. 1974. Biological control of Plant pathogens. W.H. Freeman, San Francisco.

Eswaramuthy, S. 1979. Studies on *Pseudomonas Solanacearum*. M.Sc., (Ag.) Thesis. Tamilnadu Agricultural University, Coimbatore.

Fawcett, C.H. and D.M. Spencer. 1966. Antifungal compounds in apple fruits infected with *Sclerotinia fructigena*. *Ann. Appl. Biol.*, **60**: 87-96.

Gopalsamy, G. and P. Vidyasekaran. 1987. Effect of green leaf manure on soil fertility and rice yield. *Int. Rice Res. Newsl.*, **12**: 41.

Rajan, C.P.D., and G.V. Reddy. 1986. Effect of bulky organic manures on sheath blight. *Int. Rice Res. Newsl.*, **11**: 20.

Rangasami, G. 1975. *Diseases of Crop plants in India*. Prentice - Hall of India Private Limited, New Delhi. pp. 570.

Rao, R., and M. George. 1949. Investigation of plant antibiotics 1/2 Pterogosperinin the antibacterial principle of the roots of *Marriga pterygosperma* Gaertn. *Indian J. Med. Res.*, **37**: 47-48.

Santhi, S.R. and S.P. Palaniappan. 1987. Effect of neem leaf application on nitrogen efficiency in low land rice. *Int. Rice Res. Newsl.*, **12**: 29.

Temburnikar, S.T. 1989. Effect of green manuring on the development of Bacterial Blight and brown spot. *Oryza*, **26**: 110.

Thyagarajan, P., K.M. Ramanathan, and V. Mariappan. 1986. Effect of phosphorus on bacterial blight. *Int. Rice Res. Newsl.*, **11**: 22.

Neem for the Management of Crop Diseases, 1995
Editor : V. Mariappan
Associated Publishing Co., New Delhi, India
pp. 31-.36

MANAGEMENT OF ROOT ROT AND WILT OF PIGEONPEA AND CHARCOAL ROT OF SOYBEAN BY ORGANIC AMENDMENTS

Nakkeeran, S., Kousalya Gangadharan, Jeyarajan, R. and Krishnaveni, V.

Department of Plant Pathology, Tamil Nadu Agricultural University
Coimbatore-641 003, Tamil Nadu

INTRODUCTION

Pulses have been referred to as unique jewels of Indian crop husbandry. India is a major pulse growing country in the world which share 35-36 per cent and 27-28 per cent of total area and production of pulses respectively in the world. Much emphasis has been made in recent years for the increased production of pulses and diseases are one of the constraints. In Tamil Nadu pulses are grown over an area of 8.209 L. Ha with an annual production of 3.338 L.M.T. (Agrostat '91).

Pigeonpea (*Cajanus cajan* (L.) Mill.) and soybean (*Glycine max* (L.) Merr.) are cultivated over an area of 1.25 lakh hactares and 1125 thousand hactares respectively in Tamil Nadu.

Among the various diseases affecting pigeonpea root rot (*Macrophomina phaseolina* (Tassi.) Goid and wilt (*Fusarium udum* Butler) are important. Charcoal rot of soybean (*M. phaseolina*) is also gaining importance and yield loss upto 77 per cent has been estimated (Gangopadhyay *et al.,* 1970). Two field trials were conducted for the management of these soil-borne plant pathogens by organic amendments to soil at Agricultural College and Research Institute, Coimbatore.

Dwivedi and Singh (1986) reported that neem cake, castor cake and mahua cake reduced the population of *M. phaseolina* in soil. Narasimalu and Bhaskaran (1987) opined that the addition of neem cakes and neem cake to soil reduced the sclerotial inoculum and root rot disease of soybean. Survival of *Rhizoctonia bataticola* in soil was affected by amending the soil with wheat straw and rice hull (Byadgi and Hegde, 1988). Muthusamy (1989) stated that application of organic amendments to soil, viz. neem cake, farmyard manure, neem leaves and pungam leaves decreased charcoal rot of soybean and increased the yield. Theradimani and Marimuthu (1990) reported that augumentation of the soil with decomposed coir pith curbed dry root rot of blackgram.

MATERIALS AND METHODS

Soybean: A randomised replicated trial with three replications and seven treatments for the management of charcoal rot of soybean at Tamil Nadu Agricultural University, Coimbatore during summer 1991 with Co1 cultivar was executed. The seven treatments were:

1. Farm yard manure (10 tonnes/ha)
2. Neem cake (250 kg/ha)
3. Iluppai cake (250 kg/ha)
4. *Leucaena leucocephala* leaves (10 tonnes/ha)
5. *Glyricidia maculata* leaves (10 tonnes/ha)
6. *Calotrophis gigantea* leaves (10 tonnes/ha) and
7. Control

The organic amendments were added and thoroughly mixed with field soil ten days before sowing. Germination was recorded ten days after sowing (DAS). The incidence of charcoal rot was recorded on 10, 20, 30 40, 50, 60 and 70 DAS. Yield was recorded at harvest, cost benefit ratio was worked out.

Pigeonpea: A field trial on a randomised replicated design with three replications and eight treatment was laid during kharif 1991 season to study the efficacy of organic amendments to soil on the incidence of dry root rot and wilt of pigeonpea. Co5 a susceptible cultivar was used for the studies. The eight treatments were:

1. Farm yard manure (10 tonnes/ha)
2. Decomposed coir pith (15 tonnes/ha)
3. Press mud (12.5 tonnes/ha)
4. Neem cake (150 kg/ha)
5. *Leucaena leucocephala* leaves (10 tonnes/ha)
6. Poultry manure (5 tonnes/ha)
7. Carbendazim (2g/kg) and
8. Control.

The organic amendments were applied and mixed thoroughly with the field soil, 10 days before sowing. Germination was recorded ten days after sowing (DAS). Incidence of root rot and wilt were recorded on 10, 20, 30, 40, 50, 60, 70 and 80 DAS. Yield and cost-benefit ratio were worked out.

RESULTS

Soybean: Germination was maximum (98.5%) in *G maculata* leaves amended soil and was on par with *L. leucocephala,* farm yard manure, neem cake and iluppai cake amended soil (Table 1). Neem cake recorded the least incidence of charcoal rot (15.7%) and the per cent decrease in the incidence of charcoal rot was found to be 62.4. The next best treatment being *L. leucocephala* (20.9%) while in control it was 41.8 per cent (Table 1).

Yield was maximum (1436 kg/ha) in neem cake amendment and was on par with *L. leucocephala* (1427 kg/ha) and farm yard manure (1425 kg/ha). The yield in control was only 1193 kg/ha (Table 1). Cost benefit ratio was 'maximum' in soil amended with *L. leucocephala* (1:3.7) (Table 1).

Table 1. Efficacy of organic amendments on charcoal rot incidence and yield of soybean

S. No.	Treatments	CHARCOAL ROT			YIELD		
		Germination (%)	Per cent	Per cent decrease over control	Kg/ha	Per cent increase over control	Cost benefit ratio
1.	Glyricidia maculata (10 t/ha)	98.5 (83.1)[a]	24.1 (29.4)[c]	42.3	1386[b]	16.2	1:3.0
2.	Leucaena leucocephala (10 t/ha)	98.3 (82.3)[a]	20.9 (27.2)[b]	50.0	1427[a]	19.6	1:3.7
3.	Calotropis gigantea (10 t/ha)	97.3 (80.6)[b]	25.1 (30.1)[d]	40.0	1378[b]	15.5	1:2.9
4.	Farm yard manre (10 t/ha)	98.2 (82.3)[a]	21.5 (27.7)[b]	48.6	1425[a]	19.4	1:1.4
5.	Neem cake (250 kg/ha)	98.2 (82.3)[a]	15.7 (23.3)[a]	62.4	1436[a]	20.4	1:1.4
6.	Iluppai cake (250 kg/ha)	97.6 (81.9)[a]	23.8 (29.3)[c]	43.1	1382[b]	15.8	1:0.3
7.	Control	97.3 (80.6)[b]	41.8 (40.3)[e]	-	1193	-	-
	CD (P = 0.05)	1.7	0.5		23.8		

Values in parentheses are transformed values.

Pigeonpea: Germination of pigeonpea seed was maximum (98.1%) in neem cake amendment and was on par with carbendazim (seed) application (97.8%) and decomposed coir pith amendment (97.2%). While in control the germination was 92.1 per cent (Table 2). There was a maximum reduction of wilt incidence (67.7%) by soil amendment with neem cake followed by *L. leucocephala* (64.3%), decomposed coir pith (62.6%) and farm yard manure (61.3%) amendments (Table 2). Incidence of root rot was significantly less (12.2%) in neem cake amendment and was on par with decomposed coir pith (13.0%) and *L. leucocephala* (13.2%) amendments. maximum reduction of root rot incidence (72.0%) over control was seen in neem cake amendment (Table 2).

Table 2. Efficacy of organic amendments on the incidence of wilt, root rot and yield of pigeonpea

S. No.	Treatment	WILT			ROOT ROT		YIELD		
		Germination (%)	Percent	Percent decrease over control	Percent	Percent decrease over control	Kg/ha	Percent increase over control	Cost benefit ratio
1.	*Leucaena leucocephala* (10 t/ha)	95.8 (78.5)^{b}	8.4 (16.8)^{ab}	64.3	13.2 (21.3)^{a}	69.7	913^{a}	57.1	1:1.6
2.	Farm Yard Manure (10 t/ha)	96.6 (79.5)^{b}	9.1 (17.5)^{bc}	61.3	15.9 (23.5)^{b}	63.5	842^{b}	44.9	1:1.5
3.	Decomposed coir pith (15 t/ha)	97.2 (80.5)^{ab}	8.8 (17.2)^{abc}	62.6	13.0 (21.1)^{a}	70.2	918^{a}	58.0	1:1.7
4.	Press Mud (12.5 t/ha)	94.1 (76.0)^{c}	10.2 (18.6)^{cd}	56.6	19.0 (25.8)^{c}	56.4	735^{c}	26.5	1:1.3
5.	Neem cake (250 kg/ha)	98.1 (82.7)^{a}	7.6 (16.0)^{a}	67.7	12.2 (20.5)^{a}	72.0	957^{a}	64.7	1:1.8
6.	Poultry manure (5 t/ha)	90.5 (72.1)^{d}	11.1 (19.4)^{d}	52.8	24.0 (29.3)^{d}	45.0	639^{d}	10.0	1:1.2
7.	Carbendazim (2 g/ha)	97.8 (81.6)^{a}	14.3 (22.2)^{c}	39.1	31.7 (34.3)^{c}	27.3	610^{de}	5.0	1:1.1
8.	Control	92.1 (73.8)^{cd}	23.5 (29.0)^{d}	–	43.6 (41.3)^{f}	–	581^{c}	–	–
	CD (P = 0.05)	12.3	1.4		1.35				

Values in parentheses are angular transformed values.

The yield of grains was maximum (957 kg/ha) in neem cake amendment and was on par with decomposed coir pith (918 kg/ha) and *L. leucocephala* (913 kg/ha) amendments, while in control the yield was only (581 kg/ha) (Table 2). Cost-benefit ratio was maximum (1:1.7) in neem cake and decomposed coir pith amendment (Table 2).

DISCUSSION

Charcoal rot and root rot: Charcoal rot of soybean was significantly less (15.7%) in neem cake among six organic amendments tried. In neem cake amendment there was 62.4 per cent decrease of charcoal rot of soybean over control. The yield was also found to be maximum in the same treatment (1436 kg/ha) The incidence of root rot of pigeonpea was reduced by 72.0, 70.2 and 69.7 per cent respectively in neem cake decomposed coir pith and *L. leucocephala* amendments. The yield was maximum (97.7 kg/ha) in neem cake amendment. Leguminous plant materials reduced the sclerotial population and disease caused by *M. phaseolina*

(Papavizas, 1977; Filho and Dhingro, 1980). Lakshmanan and Nair (1984) reported that the application of neem cake, groundnut cake and illuppai cake reduced the viability of sclerotia of *R. solani* upto 80 days after application. Disease caused by *M. phaseolina* in soybean was managed by the addition of organic amendments to soil (Narasimalu, 1985; Samiyappan and Arjunan, 1986 and Narasimalu and Bhaskaran, 1987). Arjunan *et al.* (1987) opined that pigeonpea root rot was reduced in alfalfa meal amended soil due to the increase in the population of *Trichoderma viride, Penicillium* sp., *Aspergillus flavus* and *Bacillus subtilis.* Byadgi and Hegde (1988) stated that soil amended with wheat straw and rice hulls reduced *R. bataticola* infection in bean seedlings. Root rot of black gram by *M. phaseolina was reduced by soil amendment with farm yard manure, neem cake, mahua cake and sugarcane press mud (Rukmani and Mariappan, 1990). Amer Singh et al.* (1990) reported that combined application of urea along with farm yard manure reduced root rot of sesamum and increased seed yield.

Wilt: There was a significant reduction in the incidence of wilt (67.7%) and increase in yield (64.7%) over control in neem cake amendment. Singh and Singh (1982) found that soil amended with oil cakes of margosa and castor was found to be inhibitory to *F. udum.* The population of *F. udum* declined in neem cake and increased activity of bacteria in the same soil (Singh and Singh, 1983). Singh *et al.* (1989) stated that soils amended with wheat, oats, chickpea and lentil residues suppressed the growth of *F. oxysporum* sp. *ciceri.* Kimura *et al.* (1991) were of the opinion that the suppression of *Fusarium* in farm yard manure amended soil was due to the production of growth suppressing substances at the specific site by gram negative bacteria.

REFERENCES

Agrostat, 1991. Directorate of agriculture. Madras-600 005. p. 28-30.

Amer Singh, T., P. Bhowmik and B.S. Choudhry, 1990. Effect of soil amendments with inorganic and organic sources of nitrogenous manures in the incidence of root rot and seed yield in sesamum. *Indian Phytopath.,* **43**: 442-443.

Arjunan, G., R. Samiyappan, N. Shanmugam and R. Jeyarajan. 1987. Control of soil borne pathogen in tropical pulses by biological means. Workshop on Biological control of plant diseases. 10-12, March, 1987, TNAU, Coimbatore. p. 33 (Abst.).

Byadgi, A.S. and R.K. Hegde. 1985. Variation among the isolates of *Rhizoctonia bataticola* from different host species. *Indian Phytopath.,* **38**: 297-301.

Dwivedi, T.S. and R.S. Singh. 1986. Survival of *Macrophomina phaseolina* from cotton in amended soil. Seminar on management of soil-borne diseases of crop plants. 8-10 Jan. 86. TNAU, Coimbatore, 5p. (Abstr.)

Filho, E.S. and O.D. Dhingra, 1980. Effect of herbicides on survival of *Macrophomina phaseolina* in soil. *Trans. Br. Mycol. Soc.,* **74**: 61-64.

Gangopadhyay, S., T.D. Wyllie and V.D. Luedders. 1970. Charcoal rot disease of soybean transmitted by seeds. *Pl. Dis. Retpr.* **54**: 1088-1091.

Kimura, M., L. Toyota and S. Kuwatsuka. 1991. Analysis of *Fusarium* suppression in farm yard manure amended soil. *Japanese Soil Sci.* and *Plant Nutrition,* **62**: 14-20.

Lakshmanan, P., and M.C. Nair. 1984. Effect of soil amendments on the viability of sclerotia of *R. solani* in soil. *Madras Agric. J.,* **71**: 526-529.

Muthusamy, S. 1986. Studies on charcoal rot of soybean (*Glycine max* (L.) Merril) caused by *Macrophomina phaseolina* (Tassi.) Goid. Ph.D. Thesis, TNAU, Coimbatore, India, pp. 127.
Narasimalu, G. 1985. Studies on the management of soybean root rot *Macrophomina phaseolina* (Tassi.) Goid. and leaf minor *Aproaerema modicella* Diventer, M.Sc., (Ag.), Thesis, TNAU, pp. 2.
Narasimalu, G. and R. Bhaskaran. 1987. Effect of organic amendments on root rot of soybean caused by *Macrophomina phaseolina. Workshop on biological Control of plant Diseases.* Dept. of Plant Pathology, TNAU, Coimbatore-3, 10-12, March 1987, p. 3. (Abst.).
Papavizas, G.C. 1977. Some factors affecting survival of sclerotia of *Macrophomina phaseolina. Can. J. Microbiol.,* **22**: 1034-1039.
Rukmani, S. and V. Mariappan. 1990. Influence of organic amendments with *Trichoderma viride* on the control of root rot of blackgram. *Pl. Dis. Res.,* **5**: 244 (Abstr.)
Samiyappan, R. and G. Arjunan. 1986. Influence of organic amendments on seed-borne fungi, germination and abnormal seedlings of soybean seeds. Seminar on management of soil-borne diseases of crop plants. 8-10 Jan., 1986. TNAU, Coimbatore, p. 30 (Abst.).
Singh, D., K.G. Neema and S.C. Vyas. 1989. Toxicity of some of the crop residues to soil-borne pathogens of chickpea. *Int. Chickpea Newsl.* **21**: 19-20.
Singh, N. and R.S. Singh. 1983. Development of *Fusarium* udum (Pigeonpea wilt pathogen) in oil cake amended soil. *Indian Bot. Reptr.* **2**: 116-120.
Singh, T. and D. Singh. 1982. Transmission of seed-borne inoculum of *M. phaseolina* from seed to plant. *Proc. Acad. Sci.,* **91**: 257-370.
Theradimani, M. and T. Marimuthu. 1990. Management of damping off in chillies on dry root rot of blackgram by soil application of decomposed coir pith. *Pl. Dis. Reptr.* **5**: 251 (Abs.).

Neem for the Management of Crop Diseases, 1995
Editor : V. Mariappan
Associated Publishing Co., New Delhi, India
pp. 37-42.

MANAGEMENT OF MAJOR FOLIAR FUNGAL DISEASES OF GROUNDNUT (*ARACHIS HYPOGAEA* L.) USING AQUEOUS LEAF EXTRACT OF NEEM (*AZADIRACHTA INDICA* A. JUSS)

Ghewande, M.P.

National Research Centre for Groundnut, Junagadh-362 001, Gujarat

Groundnut, the kingpin among the oilseed crops in India, alone accounts for 45% of the oilseeds area and 55% of the total oilseeds production. However, the average productivity is only around 900 kg/ha. An important factor contributing to low yield is attack by diseases. Among foliar diseases caused by fungi, early leaf spot (*Cercospora arachidicola* Hori), late leaf spot (*Phaeoisariopsis personata* (Berk. Curt.) v. Arx.), rust (*Puccinia arachidis* Speg.) and Alternaria leaf spot (*A. alternata* (Fr.) Keissler) are economically important. These diseases are capable of causing cumulative yield loss to the tune of 70% (Subrahmanyam *et al.* 1980; Ghewande *et al.* 1983).

Control of groundnut foliar fungal diseases has been tried mostly through the use of synthetic fungicides in the country. But repeated use of chemicals led to ecological hazards, economic problems, besides the development of resistance to some fungicides by some pathogens. Therefore, hunt for renewable sources of alternative chemotherapeutents is highly desirable. Neem appears to be the reservoir of effective chemotherapeutants and can provide renewable sources of useful biofungicides.

The chemical control of foliar diseases of groundnut through the spray of carbendazim 0.05% + mancozeb 0.2% is effective but expensive. Moreover, these chemicals sometimes are not readily available in the remote areas. As an alternative, a 2 and 5% aqueous neem-leaf extract was tried alongwith other plant leaf extracts and fungicides at the National Research Centre for Groundnut (NRCG), Junagadh to control foliar diseases during kharif, 1985, 1986, 1989 and 1990. The effect of neem leaf extract on growth, seed infection and colonization of Alternaria species was also studied under laboratory condition. The main objective of using aqueous neem leaf extract was to generate information and develop low cost protection technology within the reach of groundnut growing small and marginal farmers of the country.

MATERIALS AND METHODS

Field experiments were conducted during the rainy seasons of 1985 and 1986 with GAUG 1 and 1989 and 1990 with JL 24 varieties of groundnut highly susceptible to early leaf spot (ELS), late leaf spot (LLS) and rust at the NRCG, Junagadh. Randomized block design was followed with 4 replications in plots of 5 × 3.6 m adopting local package of practices.

Preparation of aqueous neem leaf-extract: Fresh leaves of neem (*Azadirachta indica* A. Juss.) were collected. 500 g of leaflets were washed thoroughly with water to remove dust particles and homogenized by adding little water till a fine paste was obtained. About 100 ml of water was added to the fine paste and stirred it well to get a homogenous extract. The juice was extracted from this material by squeezing through 2-3 layers of muslin cloth. The volume of the juice was made up to 1 litre by adding more water. Thus a 50% stock solution was obtained. This stock solution was used for preparing 2% and 5% aqueous neem leaf extract as given below:

— Concentration recommended - 2%
— Quantity of water required - 10 litres
— Concentration of stock solution - 50%

Hence quantity of stock solution required = 2 × 10,000 ml/50 = 400 ml. Thus, the quantity of 9,600 ml of water was taken and 400 ml of 50% stock solution of aqueous neem leaf extract was added to it to get 10 litres of 2% spray suspention. Similarly, other concentrations of aqueous neem leaf extract from 50% stock solution were prepared. Normally 450-500 litres/ha of spray suspension is required.

The aqueous neem leaf extract of 2% during kharif, 1985 and 1986 and 5% during kharif 1989 and 1990 was sprayed 3 times at 2 week intervals starting from 4-5 weeks after planting. Sprays of carbendazim 0.05% + mancozeb 0.2% and water served as the national check and control respectively. Observations on leaf spots (ELS and LLS) and rust were recorded 15 days before harvest by adopting 1-9 point scale where 1 = no disease and 9 = 50 to 100% leaf area diseased and then con verted to per cent disease intensity using the formula.

$$\text{PDI} = \frac{\sum \text{nv}}{\text{N} \times \text{Maximum index of scale}} \times 100$$

Where n = number of leaflets scored in each index value
V = index scored
N = Total number of leaflets scored

The per cent decrease in disease over control was calculated using formula:

$$\frac{\text{PDI in control} - \text{PDI in treatment}}{\text{PDI in control}} \times 100$$

Pod yield data were recorded after thorough drying of pods. The per cent increase in pod yield over control was calculated as:

$$\frac{\text{Pod yield in treatment - Pod yield in control}}{\text{Pod yield in control}} \times 100$$

Gross returns were calculated by converting the yields of groundnut into rupees by multiplying with the average market cost (Rs. 800/ha). The basic production cost (Rs. 4608/ha) was derived from the adaptive trials of All India Coordinated Project on Oilseeds (Junagadh Centre) where improved agronomic practices were compared against conventional production techniques.

The growth of Alternaria species occurring on groundnut, *A. alternata, A. tenuissima, A. raphani* was studied in aqueous neem leaf extract (1-5%). The glucose nitrate medium (Glucose-10 g; KNO_3 2.5 g; KH_2PO_4-1 g; $MgSO_4$-0.5 g; Distilled water - 1 litre) was used as control. *Alternaria* species were grown in 250 ml conical flasks on 25 ml of aqueous neem leaf extract (1-5%). Spore suspension of *Alternaria* species was prepared by adding 10 ml of sterile distilled water in 8-day-old slope cultures. Each of four flasks were inoculated with 1 ml of spore suspension of *Alternaria* species under aseptic conditions. The 4 flasks each were then incubated at 28 ± 1°C and relative humidity of 85% for 10 days. The contents of the flasks were then filtered through whatman filter paper (No. 1). The dry weight of the mycelial mat alongwith spores was recorded.

The effect of aqueous neem leaf extract on seed infection and colonization by 3 *Alternaria* species was also studied. The seeds of JL 24 variety of groundnut were soaked in 1-5% aqueous neem leaf extract for 1 hour. For control treatment, seed soaking was done in sterile distilled water. After shade drying of seeds, seeds (10) in four replications each were then placed aseptically in sterile 9 cm diameter petridishes and 1 ml each of the spore suspension of *A. alternata, A. tenuissima* and *A. raphani* were added to them. The spore suspension was spread over the seeds by gently whirling them around within the dishes. The plates were placed in a incubator at 25°C for 8 days under 98 ± 2% relative humidity. After 8 days of incubation period, observations on seed infection and colonization by *Alternaria* species were recorded.

RESULTS AND DISCUSSION

The mean data showed that the spray application of 2-5% aqueous neem leaf extract reduced early leaf spot (ELS) by 13.64% (mean of 2 years, 1989, 1990), late leaf spot (LLS) by 22.2% (mean of 4 years, 1985, 1986, 1989, 1990) and rust by 30.63% (mean of 2 years 1985, 1986) and increased pod yield by 26.1% (mean of 4 years) (over control (Table 1), with net return of Rs. 5906/ha and additional return of Rs. 2232/ha and benefit cost ratio of 2.26. The chemical treatment (carbendazim 0.05% + mancozeb 0.2%) realized reduction of ELS by 65.62% (mean of 2 years, 1989 and 1990), LLS by 35.35% (mean of 4 years), and rust by 39.13% (mean of 2 years) with increased in yield by 51.39%. This chemical treatment gave net return of Rs. 6671/ha and additional return of Rs. 4032/ha with BCR of 2.17 (Table 1).

Table 1. Effect of aqueous neem leaf extract on the severity of foliar fungal diseases and yield of groundnut and details of economics during 1985-86 & 1989-90 at the NRCG, Junagadh

Treatment	Disease reduction (%)			Pod yield** (kg/ha)	Increase in yield** (%)	Additional yield** (kg/ha)	Net return** (Rs/ha)	BCR**
	ELS*	LLS*	Rust***					
Carbendazim 0.05% + Mancozeb 0.2%	65.62	35.35	39.13	1550.29	51.39	504	6671	2.17
Aqueous neem leaf extract	13.64	22.20	30.63	1325.59	26.10	279	5906	2.26
Control (water spray)	–	–	–	1046.52	–	–	3674	1.78

* Average of two years (1989-1990)
** Average of four years (1985-1986, 1989-1990)
*** Average of two years (1985-1986)

The usefulness of aqueous neem leaf extract against foliar diseases has already been reported by Ghewande (1989), Ghewande *et al.* (1992) and was also confirmed by conducting field trials during kharif 1989 and 1990 at 5 hot spot locations (Aliyaranagar, Vriddhachalam in Tamil Nadu, Dharwad in Karnataka, Jalgaon in Maharashtra and Chiplima in Orissa) of the All India Coordinated Research Project on Oilseeds (AICORPO). The mean data of 2 years over locations proved that the aqueous neem leaf extract was efficaceous in controlling both late leaf spot and rust diseases and in increasing pod yield (Anonymous, 1991). The effective use of neem products in the control of groundnut diseases has been reported by several workers (Ganapathy and Narayanasamy, 1990; Usman *et al* 1991; Venkatarao and Alice, 1991).

The growth of *Alternaria alternata, A. tenuissima* and *A. raphani,* causing leaf spot, leaf blight and leaf blotch of groundnut was reduced by 77.84 to 95.68%, 70.02 to 95.07% and 88.52 to 92.42% respectively with 1 to 5% aqueous neem leaf extract (Table 2).

Table 2. Growth of *Alternaria* species in aqueous neem leaf extract at 28 ± 1°C after 10 days

Aqueous neem leaf extract (%)	Mycelial growth in g/l		
	A.alternata	*A.tenuissima*	*A.raphani*
1.	0.24	0.23	0.40
2.	0.26	0.40	0.48
3.	0.44	0.80	0.66
4.	0.75	0.81	0.63
5.	1.18	1.40	0.87
Control (Glucose nitrate medium)	5.55	4.32	5.28
S. Em. ±	0.18	0.12	0.10
C.D. (P = 0.05)	0.53	0.35	0.28
C.V.	21.83	15.48	12.46

The aqueous neem leaf extracts (1-5%) also inhibited seed infection by 41.19 to 100% and seed colonization by 20,04 to 100% of groundnut by these fungi (Table 3).

Table 3. Effect of aqueous neem leaf extract on seed infection and colonization by *Alternaria* species

Aqueous neem leaf extract (%)	Mean seed infection (%)			Mean seed colonization (%)		
	A.alternata	*A.tenuissima*	*A.raphani*	*A.alternata*	*A.tenuissima*	*A.raphani*
1.	33.33 (35.25)	13.33 (21.39)	26.66 (31.11)	13.33 (21.39)	0.00 (0.03)	10.00 (18.44)
2.	16.67 (24.12)	6.67 (15.00)	10.00 (18.44)	6.67 (15.00)	0.00 (0.03)	3.33 (10.47)
3.	13.33 (21.39)	0.00 (0.03)	6.66 (15.00)	3.33 (10.47)	0.00 (0.03)	3.33 (10.47)
4.	6.67 (15.00)	0.00 (0.03)	3.33 (10.47)	0.00 (0.03)	0.00 (0.03)	0.00 (0.03)
5.	3.33 (10.47)	0.00 (0.03)	0.00 (0.03)	0.00 (0.03)	0.00 (0.03)	0.00 (0.03)
Control (sterilized distilled water)	56.67 (48.85)	43.33 (41.15)	26.66 (31.11)	16.67 (24.12)	6.67 (15.00)	10.00 (18.44)
S. Em ±	10.75	6.57	9.05	7.20	2.51	5.74
C.D. at (P = 0.05)	n.s.	19.78	n.s.	n.s.	7.54	n.s.

Figures in parentheses are angular values

Based on the 4 years data generated at the National Research Centre for Groundnut, Junagadh and 2 years data of 5 locations of AICORPO, it could be concluded that although carbendazim 0.05% + mancozeb 0.2% gave better control of foliar diseases and maximum increase in pod yield, aqueous neem leaf extract was much more economical and easily adoptable at the village level. Also, neem leaf extract holds promise for potential biofungicides. This study suggests large scale use of aqueous neem leaf extract and other neem products against major foliar fungal diseases of groundnut in the country. This is a low cost plant protection technology available to groundnut growing farmers in the country.

ACKNOWLEDGEMENT

The author is grateful to Dr. P.S. Reddy, Director, National Research Centre for Groundnut, Junagadeh for facilities and encouragement.

REFERENCES

Anonymous. 1991. Annual Kharif Oilseeds Research Workers' Group Meeting, May 6-8, 1991 GAU, Junagadh, Directorate of Oilseeds Research, Rajendranagar, Hyderabad.

Ganapathy, T. and P. Narayanasamy. 1990. Effect of Plant products on the incidence of major diseases of groundnut. *Int. Arachis Newsl.* **7**: 20-21.

Ghewande, M.P. 1989. Management of foliar diseases of groundnut (*Arachis hypogaea* L.) using plant extracts. *Indian J. Agric. Sci.,* **59**: 133-134.

Ghewande, M.P., S. Desai, Prem Narayan and A.P. Ingle. 1992. Evaluation of some new fungicides and plant products to manage foliar diseases of groundnut (*Arachis hypoqaea* L.). Proceedings, 'Chemical Management of Plant Pathogens' of 4th Annual Zonal Meeting (WZ) of Indian Phytopathological society, M.U. Aurangabad (1-2 Nov. 1991) (In Press).

Ghewande, M.P., A.K. Shukla, R.N. Pandey, and D.P. Misra. 1983. Losses in groundnut yields due to leaf spots and rust at different intensity leavels. *Indian J. Mycol. Pl. Pathol.,* **13**: 125-127.

Subrahmanyam, P., V.K. Mehan, D.J. Nevill and D. Mc Donald. 1980. Research on fungal diseases of groundnut at ICRISAT. *In*: ICRISAT (International Crops Research Institute for the Semi-Arid Tropics), 1980. Proc. 1st Intl. Workshop on groundnut, 13-17 October, 1980, Patancheru, A.P., India, pp. 193-198.

Usman, K.M., P. Durairaj, K. Karunanithi and S. Thangavelu. 1991. Effect of neem products on the control of foliar diseases of groundnut. Poster Abstract, Second International Groundnut Workshop, 25-29, November, 1991. ICRISAT, Patancheru, A.P., India.

Venkatarao, A. and D. Alice. 1991. Neem extract checks plant diseases. *Capart Press Clippings,* **7**: 23.

Neem for the Management of Crop Diseases, 1995
Editor : V. Mariappan
Associated Publishing Co., New Delhi, India
pp. 43-47.

USE OF NEEM PRODUCTS FOR THE MANAGEMENT OF BASAL STEM ROT DISEASE OF COCONUT

Bhaskaran, R.

Coconut Research Station, Tamil Nadu Agricultural University
Veppankulam-614 906, Thanjavur District, Tamil Nadu

Coconut (*Cocos nucifera* L.) palm is affected by more than fifty diseases in different parts of the world of which some of them are lethal and many of them are debilitating in nature. In Tamil Nadu, basal stem rot disease (BSR) caused by *Ganoderma lucidum* (Leys) Karst. is the most serious disease affecting coconut production to a considerable extent. This disease is also called as Thanjavur wilt, *Ganoderma* disease, *anabe* disease and bole rot in different states in India.

In Tamil Nadu, this disease was first observed in coconut in Thanjavur district in 1952. Characteristic symptoms of the disease are exudation of reddish brown viscous fluid from the basal portions of the stem which gradually extend upwards, drooping, drying and falling of leaves, extensive root rotting and death of the palm (Bhaskaran *et al.,* 1989).

Lack of soil moisture and increase in soil temperature increase the severity of the disease. The disease spreads fast in sandy soil and in soils poor in organic matter.

Soil drenching with 40 litres of 1 per cent Bordeaux mixture and stem injection of Aureofungin-sol 2 g with 1 g of copper sulphate in 100 ml of water thrice at quarterly interval is the control measure recommended to reduce the severity of the disease and to increase the nut yield (Bhaskaran and Ramanathan, 1982; Bhaskaran *et al.,* 1984). However, the palms are not completely cured of this disease by the fungicides treatment and the treatment schedule should be repeated once in three years (Ramadoss and Bhaskaran, 1987). Thus the chemical treatment becomes costly and hence the use of neem products for the management of basal stem rot affected coconut assumes significance. In the present paper, the work done so far on the effect of neem products in the management of BSR are reviewed.

Effect of neem products on growth of *Ganoderma lucidum*: To find out the effect of neem products on the *in vitro* growth of *G. lucidum,* neem cake extract and neem seed kernel extract were tested at different concentrations. Neem cake extract at 10 per cent concentration completely inhibited the mycelial growth of *G. lucidum* (Bhaskaran *et al.,* 1988). Neem seed kernel extract at 5 per cent concentration gave 34 per cent inhibition of mycelial growth.

Effect of neem cake on disease intensity: Efficacy of neem cake in comparison with other organic manures and Bordeaux mixture was tested for the control of BSR. Among the organic manures tested, neem cake was found to be more effective either alone or in combination with Bordeaux mixture soil drenching when compared to other treatments. Neem cake treatment also gave significantly higher yield than control (Table-1).

Table 1. Effect of organic manures on basal stem rot intensity and nut yield in coconut (Bhaskaran and Ramanathan, 1983)

Treatments	Disease index	Mean Nut yield* per palm
Neem cake 5 kg (NC)	12.3	39
NC + 1% Bordeaux mixture 40 litres as soil drench (NC + BM)	11.4	50
Tank silt (300 kg) (TS)	55.8	31
TS + BM	41.5	36
Farm yard manure 200 kg (FYM)	56.5	42
FYM + BM	61.9	45
Green leaves 50 kg (GL)	114.1	20
GL + BM	35.2	33
BM	69.8	30
Control	117.7	19
C.D. (P = 0.05)	38.7	7

* Mean of four years yield (1978-'81).

Gunasekaran *et al* (1986) reported that application of 10 kg of neem cake per palm per year was as effective as Bordeaux mixture soil drenching combined with aureofungin-sol stem injection.

In a study conducted at Coconut Research Station, Veppankulam, efficacy of 5 kg and 10 kg of neem cake application with Aureofungin-sol root feeding + Bordeaux mixture soil drenching + neem cake (AF + BM + NC) was compared (Bhaskaran *et al.,* 1988). It was found that 5 kg of neem cake was as effective as 10 kg of neem cake. Though the disease intensity in NC applied palms is slightly higher than in AF + BM + NC treatment, the disease intensity was very low when compared to control palms (Table-2).

Efficacy of neem cake in the management of basal stem rot disease (Bhaskaran *et al.,* 1988)

Treatments	Disease index	
	1985	1986
Neem cake 5 kg per palm (NC)	9.4	12.2
Neem cake 10 kg per palm	9.1	12.1
AF + BM + NC	8.2	10.6
Control	23.9	63.0
C.D. (P = 0.05)	0.6	1.3

Effect of neem cake on soil microbial population: Populations of fungi, bacteria, actinomycetes and *Trichoderma* were estimated in neem cake applied soil by serial dilution technique using rose-bengal agar for fungi, soil extract agar for bacteria, Kuster's agar for actinomycetes and the special medium of Elad and Chet (1983) for *Trichoderma*. Observations recorded indicated that the population of fungi and bacteria were stimulated significantly in neem cake applied soil when compared to control palms. There was significant increase in the population of *Trichoderma* also due to neem cake application (Table-3).

Table 3. Effect of neem cake on the microbial population* in soil (Bhaskaran *et al.*, 1988)

Treatments	Fungi (10^3)		Bacteria (10^5)		Actinomycetes (10^2)		*Trichoderma* (10^2)	
	A	B	A	B	A	B	A	B
Neem cake 5 kg (NC)	44	178	18	190	11	21	9	72
Neem cake 10 kg	36	173	18	182	13	26	6	64
AF + BM + NC	41	72	20	175	16	24	11	22
Control	39	40	17	19	12	15	8	6
C.D. (P = 0.05)	N.S.	11	N.S.	21	N.S.	6	N.S.	8

* Population per g of air dry soil
A Initial population;
B One month after treatment.

Survival of *Trichoderma* in neem cake and other organic amendments: In the preliminary experiments, *Trichoderma harzianum* was found to be antagonistic to *G. lucidum*. The population of *Trichoderma* increased by the application of neem cake. Hence to find out the survival of *Trichoderma* in different organic amendments, a pot culture experiment was conducted. *T. harzianum* multiplied in sterilized farm yard manure was added to the sterilized soil with different organic amendments. Population of *Trichoderma* estimated one year after initiation of the experiment showed that neem cake is the best amendment for sustaining the maximum population level of *Trichoderma* (Table-4).

Table 4. Survival of *T. harzianum* in different organic amendments

Treatments	Dosage per pot (g)	Population (10^7 cfu/g)
Neem cake	30	36
Farm yard manure	300	22
Coir dust	300	4
Pleurotus treated coir dust	300	16
Poultry manure	60	8
Tank silt	600	12
Green leaves	300	5
Control	–	3
C.D. (P = 0.05)		5

Efficacy of neem cake with *T. harzianum* in BSR control: To find out the efficacy of neem cake in sustaining the population level of *Trichoderma* under field conditions and in the management of BSR another field experiment was conducted, in which *T. harzianum* was multiplied in different organic amendments and applied to the diseased trees. Disease intensity recorded one year after application of *T. harzianum* with different organic amendments, showed that the antagonist applied with neem cake is the best treatment for the control of BSR giving a disease index of 23.6 as compared to 65.1 in control (Table-5).

Table 5. Effect of *Trichoderma harzianum* in different organic amendments on basal stem rot disease intensity in coconut and soil microbial population

Treatments	Disease Index	Soil microbial population							
		Initial				One year after treatment			
		F	B	A	T+	F	B	A	T
T.h. + Neem cake 5 kg	23.6	14	12	11	52	18	15	14	45
T.h. + Farm yard manure 50 kg (FYM)	40.5	13	14	15	44	20	22	18	40
T.h. + Tank silt 100 kg	56.6	10	16	12	38	12	18	16	34
T.h. + Coir dust 50 kg	86.8	9	12	10	43	12	14	12	38
T.h. + *Pleurotus* treated coir dust 50 kg	49.1	9	15	11	37	11	15	12	30
T.h. + Poultry manure 10 kg	68.4	11	17	13	40	13	16	13	32
Bordeaux mixture 1% 40 litres/ palm (BM)	43.0	1	13	10	1	2	15	12	0
T.h. in FYM + BM after one month	26.1	14	19	22	46	16	23	12	38
BM + FYM with *T.h.* after one month	37.6	2	12	14	42*	24	25	14	42
Control	65.1	2	13	12	0	1	15	14	0
C.D. (P = 0.05)	5.7								

F - Fungi 10^3 cfu/g of soil, B - Bacteria 10^6 cfu/g; A - Actinomycetes 10^3 cfu/g, T^+ - *Trichoderma* 10^5 cfu/g and T - *Trichoderma* 10^3 cfu/g.
* Population estimated immediately after FYM (with *Trichoderma*) application.

Microbial population in the rhizosphere of *T. harzianum* (with different organic amendments) applied trees was estimated. In general, microbial population was high in neem cake and farm yard manure treatments when compared to control. Population of *Trichoderma* decreased in all the treatments, one year after organic manures application. The population decreased when compared to the initial population but still, *Trichoderma* population in the organic manures applied tree basins was much higher than control (Table-5).

Effect of NSKE on basal stem rot disease: To test the efficacy of NSKE (5%) on BSR, 100, 200 and 400 ml quantities of the extract were fed through roots once in three months for one year. Five replications were maintained for each treatment. Disease intensity recorded one year after the treatment showed that root feeding of 400 ml of NSKE is better than control but not as effective as annual application of 5 kg of neem cake (Table-6).

Table 6. Effect of NSKE root feeding on basal stem rot disease of coconut

Treatments	Disease index
Root feeding of NSKE (5%) 100 ml	58.6
Root feeding of NSKE (5%) 200 ml	64.0
Root feeding of NSKE (5%) 400 ml	36.0
Neem cake 5 kg	22.8
Control	63.4

There are several reports on the efficacy of neem products in the control of soil-borne diseases of crop plants (Maiti *et al.*, 1987; Padmakumari and Balakrishnan, 1987). Disease suppression may be due to the increase in the population of antagonistic microflora in soil or due to the direct effect of the organic amendment on the pathogen. In the case of BSR of coconut, neem cake was found to completely inhibit the growth of *G. lucidum* under *in vitro* and the population of *Trichoderma* in soil significantly increased due to neem cake application with a parallel decrease in disease severity. Hence use of neem cake offers great scope for the management of BSR of coconut.

REFERENCES

Bhaskaran, R., N. Ramadoss and T.K. Ramachandran. 1988. Biological control of Thanjavur wilt disease of coconut. *Indian Coconut J.*, **19**(6): 1-6.

Bhaskaran, R. and T. Ramanathan. 1982. Management of Thanjavur wilt of coconut. *Proceedings of the Fifth Annual Symposium on Plantation Crops*, CPCRI, Kasaragod. pp. 521-526.

Bhaskaran, R. and T. Ramanathan. 1983. Role of fertilizers and organic manures in Thanjavur wilt of coconut. *Indian Coconut J.*, **14**(3): 1-5.

Bhaskaran, R., T. Ramanathan and M. Ramaiah. 1984. Thanjavur wilt of coconut; its occurrence, spread and management. *Proceedings of the Sixth Annual Symposium on Plantation Crops*, CPCRI, Kasaragod. pp. 107-113.

Bhaskaran, R., P. Rethinam and K.K.N. Nambiar. 1989. Thanjavur wilt of coconut. *J. Plant. Crops.*, **17**: 69-79.

Elad, Y. and I. Chet. 1983. Improved selective media for isolation of *Trichoderma* spp or *Fusarium* spp. *Phytoparasitica*, **11**: 55-58.

Gunasekaran, M., N. Ramadoss, M. Ramiah, R. Bhaskaran, and T. Ramanathan. 1986. Role of neem cake in the control of Thanjavur wilt of coconut. *Indian Coconut J.*, **17**(1): 1-6.

Maiti, D., D.C. Khatua, and C. Sen. 1987. Effect of oil cakes and a potential antagonist on survival of sclerotia of *Sclerotium rolfsii* and stem rot of groundnut. *Proceedings of the workshop on Biological control of Plant Diseases*, Tamil Nadu Agrl. Univ., Coimbatore. pp. 1. (Abst.).

Padmakumari, G. and S. Balakrishnan. 1987. Effect of organic amendments on the survival of *Rhizoctonia solani*, the sheath blight pathogen. *Proceedings of the Workshop on Biological control of Plant Diseases*, Tamil Nadu Agrl. Univ., Coimbatore. pp. 4. (Abst.).

Ramadoss, N., and R. Bhaskaran. 1987. Effect of fungicides in the management of Thanjavur wilt. *International symposium on Ganoderma wilt Diseases on Palms and other Perennial Crops*, Tamil Nadu Agrl. Univ., Coimbatore. pp. 12-13. (Abst.).

Neem for the Management of Crop Diseases,.1995
Editor : V. Mariappan
Associated Publishing Co., New Delhi, India
pp. 49-51.

ANTIFUNGAL ACTIVITY OF NEEM (*AZADIRACHTA INDICA*) LEAVES AGAINST SOME PHYTOPATHOGENIC FUNGI

Datar, V.V.

College of Agriculture, Latur-413 512, Maharastra State, India.

From the very early times plant pathologists, following the lead given by scientists of medicine, have exhibited great interest in the use of plnat products, for control of plant diseases. Neem (*Azadirachta indica*) is known to have medicinal, and insect repellant activity due to the presence of the alkaloid azadirachtin. In the present investigations extract of neem leaves was tested for its efficacy on mycelial growth, and conidial germination of phytopathogenic fungi.

MATERIALS AND METHODS

Six fungi viz. *Curvularia lunata, Alternaria tenuis, Fusarium* sp., *Sclerotium rolfsii, Macrophomina phaseolina* and *Elsinoe amplina* were selected as test fungi for investigating the antifungal activity of neem leaf extract. These fungi were isolated from their respective host and maintained on PDA. Neem leaf extract was prepared by following procedure as described by Bambode and Shukla (1973).

Two hundred gms of leaves were washed in distilled water. They were chopped into small pieces and grinded with pestle and mortar. This material was boiled in 1 litre of distilled water for 20 minutes. The extract was strained through muslin cloth. Final quantity of extract was maintained at 500 ml. This extract was used for incorporation into PDA and also for conidial germination test.

This extract was mixed with 500 ml of freshly prepared double strength PDA (potatoes 200g, dextrose 20g, agar 20 g and distilled water 500 ml). Final quantity of leaf extract plus PDA was maintained at 1 litre. Single strength PDA without plant extract served as control. The medium was sterilized and 20 ml poured per plate aseptically. Five mm mycelial disc from actively growing cultures were used for inoculation. Observations on colony diameter were recorded periodically. Per cent reduction in growth was tabulated with reference to growth of fungi in control plates.

For testing the efficacy of neem leaf extract on conidial germination, conidia harvested from seven day old culture were suspended in distilled water to give 70-80 conidia per field of microscope (10x10x). One drop of conidial suspension was placed gently on the coverslip. A drop of plant extract was placed on this. This coverslip was placed in inverted position in cavity slide, which in turn was kept in moist chamber at 28 ± 2°C. Observation on per cent germination were recorded after three hours and six hours.

RESULTS AND DISCUSSION

The percent reduction in mycelial growth and conidial germination of six pathogenic fungi is given in Table 1.

Table 1. Efficacy of neem leaf extract on the mycelial growth and conidial germination of phytopathogenic fungi

Phytopathogenic fungi	Percent reduction in			
	Mycelial Growth		Conidial Germination	
	4 days	8 days	3 hours	6 hours
1. *Fusarium* sp.	7.5	11.95	26.72	28.13
2. *Curvularia lunata*	11.50	13.00	40.70	41.08
3. *Alternaria tenuis*	45.10	50.00	50.32	43.78
4. *Sclerotium rolfsii*	9.02	16.22	–	–
5. *Macrophomina phaseolina*	5.27	9.47	–	–
6. *Elsinoe ampelina*	37.42	39.62	39.12	37.0

The results indicated that neem leaf extract reduced the mycelial growth of all test fungi after four and eight days. After four days incubation, maximum inhibition of 45.1% was found in *Alternaria tenuis* while least inhibition of 5.2 was observed in *Macrophomina phaseolina.* After 8 days of incubation maximum inhibition of 50% was observed in *Alternaria tenuis* and minimum inhibition of 9.47% was observed in *Macrophomina phaseolina.*

Conidial germination of *Alternaria tenuis* was reduced to the extent of 50.32% after 3 hrs incubation in neem leaf extract, while lowest reduction in germination count of 26.72% was observed in *Fusarium* sp. After 6 hrs incubation. Also maximum conidial germination of 43.78% was observed in *Alternaria tenuis* while lowest inhibition of conidial germination was observed in *Fusarium* sp. Khan *et al.* (1974) reported that extract from oil cake of *Azadirachta indica* incorporated in the culture medium, inhibited the growth of all phytopathogenic fungi tested. Growth reduction was proportional to the concentration used. Bhowmick and Choudhary (1982) found that in test with 10 plant spp. greatest inhibition *in vitro* was obtained by *Acalypha indica* followed by *Camphora* sp., *Vitex negundo,* and *Azadirachta indica.*

REFERENCES

Bambode, R.S. and V.N. Shukla. 1973. Antifungal properties of certain plant extracts against some fungi. *P.K.V. Res. J.*, **2**(1): 1-8.

Bhowmick, B.N. and B.K. Choudhary. 1982. Antifungal activity of leaf extracts of medicinal plants on *Alternaria alternata* (fr.) keissler. *Indian Bot. Reptr.,* **1**(2): 164-165.

Khan, M.N., M.M. Alam, A.M. Khan and S.K. Saxena. 1974. Effect of water soluble fractions of Oil cakes and bitter principles of neem on some fungi and nematodes. *Acta Bot. Indica,* **2**(2): 120-128.

Neem for the Management of Crop Diseases, 1995
Editor : V. Mariappan
Associated Publishing Co., New Delhi, India
pp. 53-61.

MANAGEMENT OF DISEASES OF VEGETABLE CROPS BY USING NEEM PRODUCTS IN NILGIRIS DISTRICT, TAMIL NADU

Mohan, L., Mariappan, V. and Yesuraja, I.

Department of Plant Pathology, Tamil Nadu Agricultural University
Coimbatore-641 003, Tamil Nadu

The temperate vegetables like potato, cabbage, spices like garlic, ginger and others are cultivated in Nilgiris district all round the year in three seasons. These crops are invariably affected by many diseases causing significant loss. Continuous cropping and the favourable weather conditions prevailing at Nilgiris have been found favourable for the perpetuation of the pathogens and their spread. The farmers have been trying their level best to protect their crop from the pests and diseases by applying heavy doses of toxic chemicals like insecticides, fungicides etc. Annually about 26 metric tonnes of granules and thirty thousand litres of emulsifiable concentrates of various chemicals are put into horticultural ecosystems at Nilgiris. Although these chemicals could afford excellent control of insect pests and pathogens their toxicity to non-target organisms, environmental pollution, residue in the consumable products and development of resistance by the insect pests and phytopathogenic populations are now posing greater concern.

Plant products from neem have been reported to be highly effective against target insect pests and pathogens of crop plants in recent years. They were also reported to be suitable to the environmental saftey as they are quickly biodegradable. Hence, the neem preparations were tested to assess their efficacy against some of the important diseases of vegetable crops at Nilgiris.

Among the diseases that cause damage to the vegetable crops leaf roll virus disease of potato, late blight disease of potato (*Phytophthora infestans* (Mont.) de Barry), Mycosphaerella leaf spot of cabbage (*Mycosphaerella brassicola* (Fr. ex Duby) Lindau., blast disease of 'garlic' (*Botrytis allii* Munn.), and phyllosticta leaf spot of 'ginger' (*Phyllosticta gingibari*) are of considerable importance as they have been causing severe loss almost in all seasons. The preparations from neem (neem oil and neem seed kernel extracts) were tested at 0.4 and 4.0 per cent concentrations respectively in field in comparison with the standard recommended fungicides individually for each disease during 1993-'94.

Neem oil emulsion

Materials: Neem oil 4 l, water 1000 l and liquid detergent (Teepol) 1 l. The liquid detergent was dissolved in water and to this mixture neem oil was added slowly. The content was mixed and made into a milky emulsion by constant stirring.

Neem seed kernel extract: Neem seed kernel 4 kg, water 100 l, and liquid detergent 1 l.

The neem seed kernel was pounded into powder in a mortar and taken in a gunny bag and the bag was immersed in a bucket containing 20 l of water for eight hours. Then the bag was removed and the volume of the liquid was made upto 100 l by adding the remaining quantity of water. To this neem seed kernel extract, liquid detergent was added and the mixture was used for spraying the field.

The two preparations were sprayed at 500 l/ha by using a high volume sprayer.

Management of potato leaf roll virus: The effect of neem oil in comparison with application of insecticides was tested for the control of leaf roll virus transmitted by aphid vector in a field trial. Carbofuran was applied (at the rate of 2.5 kg a.i/ha) to all plots except for control plots as basal application before planting. For one set of plots carbofuran application to soil was repeated after one month. For the remaining plots insecticides (monocrotophos 0.04 per cent, endosulphan 0.07 per cent, fenvalerate 0.01 per cent, quinalphos 0.05 per cent, acephate 0.065 per cent, cartap hydrochloride 0.05 per cent, triazophos 0.08 per cent, chlordane 0.04 per cent, chlorpyriphos 0.04 per cent) and neem oil 0.4 per cent were applied separately as individual treatments. The spraying of insecticides and neem oil was given for eight times at weekly interval starting from a fortnight after planting. Control plots were not applied with any chemical either as basal treatment or as foliar spray. The per cent incidence of leaf roll virus infected plants was worked by examining the total population of the plants from individual treatments and the yield per ha also was assessed (Table 1).

Table 1. Effect of neem oil and insecticides on the incidence of potato leaf roll virus disease

Sl. No.	Treatments		Leaf roll incidence (%)	Yield (Tonnes/ha)
1.	Carbofuran 2.5 kg ai/ha	+ Monocrotophos 0.04%	7.00ab	15.3ab
2.	"	+ Endosulfan 0.07%	4.44a	14.4ab
3.	"	+ Fenvalerate 0.01%	9.33ab	16.8a
4.	"	+ Quinalphos 0.05%	6.21ab	11.5bc
5.	"	+ Acephate 0.065%	9.10ab	13.3ab
6.	"	+ Cartaphydrochloride 0.05%	7.33ab	14.0ab
7.	"	+ Triazophos 0.08%	6.88ab	14.0ab
8.	"	+ Chlordane 0.04%	9.28ab	13.9ab
9.	"	+ Chlorpyriphos 0.04%	4.44a	14.6ab
10.	"	+ Neem oil 0.4%	5.77a	14.6ab
11.	"	+ Carbofuran 2.25 kg ai/ha	4.22a	15.9ab
12.	Control (no treatment)		11.55b	6.7c

(Annual report of HRA, Vijayanagaram, 1992)

The results indicated that all the treatments were effective in reducing the disease severity and also in increasing the yield significantly when compared to control. Among the treatments carbofuran + endosulphan, carbofuran + chlorpyriphos, carbofuran + carbofuran and carbofuran + neem oil recorded the least incidence of the disease and were on par. The treatment carbofuran + fenvalerate recorded the maximum yield of 16800 kg/ha followed by carbofuran + neem oil, carbofuran + chlorpyriphos and carbofuran + endosulphan which recorded 14583, 14583 and 14441 kg/ha respectively and they were on par. The rest of the other treatments also increased the yield while control plots recorded 6,666 kg/ha only.

*Management of potato late blight (*Phytophthora infestans *(Mont.)* de Barry): Field trials were conducted with seven treatments: neem oil 0.4 per cent, neem seed kernel extract 4.0 per cent, chlorothalonil (Tetra chloroisophthalo-nitrile) 0.2 per cent, metalaxyl MZ (Methyl-DL-N-2, 6-dimethyl phenyl-N-2 methoxy acetyl alaninate) 0.25 per cent, copper oxychloride 0.25 per cent, mancozeb (Manganous-Zinc ethylene bis dithiocarbamate) 0.25 per cent and control. Six replications were maintained. The treatments were given at weekly intervals for six times starting from 20 DAS using 500 l/ha. The disease severity was scored following 0-9 scale and the yield data were recorded (Table 2). Among the treatments fungicides were relatively more effective in reducing the late blight disease than that of the neem preparations.

Table 2. Effect of neem oil and neem seed kernel extract on the late blight of potato

Sl. No.	Treatments	Mean disease index	Yield (Tonnes/ha)
1.	Neem oil 0.4%	8.3	2.40
2.	Neem seed kernel extract 4%	8.1	2.25
3.	Chlorothalonil 0.2%	2.6	11.58
4.	Metalaxyl MZ 0.25%	1.0	12.66
5.	Copper oxychloride 0.25%	5.8	5.78
6.	Mancozeb 0.25%	2.1	8.80
7.	Control	8.3	2.03
	CD (P = 0.05)	0.568	0.470

Mycosphaerella leaf spot of cabbage: The leaf spot of cabbage caused by *Mycosphaerella brassicola* (Fr. ex Duby) Lindau. is a serious disease causing considerable loss in yield.

Field trials were conducted with eight treatments including neem and replicated for four times. The preparations were applied for four times. The preparations were applied for six times at weekly interval starting from 30 DAS. The treatments were, neem oil 0.4 per cent, neem seed kernel extract 4.0 per cent, mancozeb 0.25 per cent, chlorothalonil 0.20 per cent, copper oxychloride 0.25 per cent, thiophanate methyl 0.1 per cent, carbendazim 0.1 per cent and control. The disease severity was scored using 1-9 scale and mean disease index was worked out. The results are presented in table 3. All the treatments of fungicides and neem preparations were effective in reducing the disease incidence significantly when compared to control. The yield recorded by these treatments also were consierably higher than the yield obtained from control plots. Carbendazim has recorded the maximum yield of 8.35 t/ha as against 6.17 t/ha in control. This was followed by chlorothalonil, neem oil and neem seed kernel extract which recorded 8.00, 7.97 and 7.90 t/ha respectively (Table 3).

Table 3. Effect of neem oil and neem seed kernel extract on the *Mycosphaerella* leaf spot of cabbage

Sl. No.	Treatments	Mean disease index	Yield Tonnes/ha
1.	Neem oil 0.4%	2.45	7.97
2.	Neem seed kernel extract 4%	2.25	7.90
3.	Mancozeb 0.25%	2.63	7.87
4.	Chlorothalonil 0.20%	2.19	8.00
5.	Copper oxychloride 0.25%	3.28	7.60
6.	Thiophanate methyl 0.1%	2.25	7.35
7.	Carbendazim 0.1%	1.92	8.35
8.	Control	4.53	6.17
	CD (P = 0.05)	0.193	0.345

Blast disease of garlic (*Allium sativum* L.): The blast disease of garlic caused by *Botrytis allii* Munn. has been a devastating disease in Nilgiris. The disease is controlled by spraying fungicides by the farmers. In order to assess the efficacy of neem preparations a field trial has been conducted along with the usual fungicides using garlic variety Ooty 1. A total of nine treatments were maintained in four replications. The treatments included neem oil 0.4 per cent, neem seed kernel extract 4.0 per cent, mancozeb 0.25 per cent, carbendazim 0.1%, thiophanate methyl 0.1 per cent, copper oxychloride 0.2%, metalaxyl MZ 0.25 per cent, chlorothalonil 0.2 per cent and control (no treatment). The neem preparations were applied at weekly intervals for six times starting from 30 DAS. The fungicides were applied at forthightly intervals for three times. The disease severity was graded in 1-9 scale at the end of the experiment and the mean disease index has been worked out. The yield data was worked out. From the results it was evident that the disease severity was reduced significantly with a corresponding increase in yield when compared to control. Among the treatments thiophanate methyl, chlorothalonil, neem oil, carbendazim and neem seed kernel extract were significantly more effective than the other treatments in reducing the disease severity. The treatments chlorothalonil, mancozeb, thiophanate methyl, neem oil and neem seed kernel extract were comparable in their efficacy with the fungicides and also recorded higher yields similar to the yield obtained from plots treated with fungicides (Table 4).

Leaf spot disease of ginger: Ginger (*Zingiber officinale* Roscoe) crop is affected by leaf spot caused by *Phyllosticta gingibari* cause intense spotting on foliage and the foliage dry up prematurely. Whenever the disease occur in severe form the

Table 4. Effect of neem oil and neem seed kernel extract on the blast disease of garlic

Sl. No.	Treatments	Mean disease index	Mean yield Tonnes/ha
1.	Neem oil 0.4%	1.49	3.53
2.	Neem seed kernel extract 4%	1.83	3.47
3.	Mancozeb 0.25%	2.20	3.83
4.	Carbendazim 0.1%	1.69	3.48
5.	Thiophanate methyl 0.1%	0.96	3.65
6.	Copper oxychloride 0.2%	2.10	2.88
7.	Metalaxyl MZ 0.25%	3.97	2.44
8.	Chlorothalonil 0.2%	1.07	3.96
9.	Control	5.73	1.98
	CD (P = 0.05)	0.405	0.172

productivity of the crop yield reduce considerably. Spraying fungicides have been found to be effective. The efficacy of neem products in comparison with the fungicides has been evaluated by conducting a field trial. The treatments tested were neem oil 0.4 per cent, neem seed kernel extract 4.0 per cent, chlorothalonil 0.2 per cent, mancozeb 0.25 per cent, copper oxychloride 0.25 per cent, carbendazim 0.1 per cent, thiophanate methyl 0.1 per cent and control. the treatments were replicated for four times and variety Rio-de-genero was used. The treatments were given 10 DAS and neem preparations were sprayed for four times at 10 DAS and neem preparations were sprayed for four times at 10 days interval while the fungicides were sprayed twice at 20 days interval. The disease severity was graded using 1-9 scale and the mean disease index has been worked out (Table 5).

Table 5. Effect of neem oil and neem seed kernel extract on the leaf spot of ginger

Sl. No.	Treatments	Mean disease index	Mean yield Tonnes/ha
1.	Neem oil 0.4%	1.82	4.42
2.	Neem seed kernel extract 4%	1.91	4.43
3.	Chlorothalonil 0.2%	1.00	4.62
4.	Mancozeb 0.25%	2.43	4.32
5.	Copper oxychloride 0.25%	2.43	4.35
6.	Carbendazim 0.1%	1.98	4.53
7.	Thiophanate methyl 0.1%	2.08	4.34
8.	Control	4.58	2.72
	CD (P = 0.05)	0.393	0.215

From the results obtained all the treatments were found to be effective in reducing the disease with a corresponding increased yield than control. The fungicide chlorothalonil has been highly effective in reducing the disease severity ot a minimum extent. Neem preparations also were as effective as chlorothalonil. Chlorothalonil and carbendazim treated plots recorded the maximum yield of 4.62 and 4.53 t/ha respectively. Neem oil and neem seed kernel extract treated plots recorded the average yield of 4.42 and 4.43 t/ha respectively. The neem preparations were as effective as fungicide in reducing ginger phyllosticta leaf spot and also increased the yield significantly.

DISCUSSIONS

Vegetable crops are grown all round the year and they provide a continuous source income to the farmers of Nilgiris district. But all of them crops are invariably damaged by some pests and diseases continuously as the prevailing weather conditions and the continuous cropping are highly favourable for their perpetuation. The farmers in this district usually apply pesticides more frequently than the farmers of other districts and the practice of indescriminate applications are not only adding high input cost but also load to the undesirable results to the environment. In recent years several plant products have been demonstrated to be useful for the management of pests and diseases of crop plants. Pesticides of plant origin not only afford protection to the crop plants from pests and diseases but also reported to leave least or no residue due to their quick biodegradable nature.

The efficacy of seed oils in reducing the virus diseases of crop plants has been reported in several instances. The transmission of chilli mosaic virus to chilli plants that were treated with neem oil (*A. indica*) and mahua oil (*Madhuca longifolia*) at 5 per cent concentration three hours prior to the inoculation feeding by the aphid vectors *Aphis craccivora* and *Myzus persicae* has been reported to be reduced to an extent of 53.33 to 73.3 per cent (Duleep Kumar Samuel, 1990).

The whitefly (*Bemisia tabaci*) transmitted mungbean yellow mosaic virus incidence was reduced when the crop was sprayed with seed oils of neem (*A. indica*) pungam (*Pongamia pinnata*) and cotton (*Gossypium* sp.) at 0.1 and 0.5 per cent concentrations 15, 22, and 25 days after sowing. Neem oil at both concentrations recorded 23.06 and 22.38 per cent incidence respectively as against 34.81 per cent in control. The plants treated with the neem oil at both concentrations recorded 567 and 558 kg/ha while the yield obtained from control plot was only 403 kg/ha. Neem oil has been more effective than pungam oil or cotton seed oil (Ravindra Babu, 1987).

Mariappan *et al.* (1988a) reported that the nonedible oil extracted from seeds of Karanji (*Pongamia pinnata*), mahua (*Madhuca longifolia* var. *latifolia*) and punnai (*Calophyllum inophyllum*) were as effective as neem (*A. indica*) in reducing the transmission of rice tungro virus by the green leaf hopper (*Nephotettix virescens*).

In the present study the leaf roll virus disease of potato transmitted by the aphid (*Myzus persicae*) has been reduced effectively in plots treated with carbofuran (2.5 kg a.i./ha as soil application before sowing followed by foliar spray with neem oil. The leaf roll virus incidence in plots sprayed with neem oil was 5.77 per cent only and it was comparable with the other treatments involving insecticides. In control plots the incidence was high (11.55 per cent). The yield recorded from plots treated with neem oil was 14.6 t/ha and was on par with the yield recorded from insecticide treated plots and was higher than the 6.7 t/ha recorded from control plots.

Thus the neem oil spray has been found to be as effective as insecticides in reducing the leaf roll virus disease of potato.

The effectiveness of neem oil in reducing the fungal diseases have been well demonstrated. (Chaturvedi *et al.*, 1987) reported that an essential oil isolated from *Adenocalynima alliaceum* was more active than some commercial fungicides tested and was able to prevent brown spot disease of rice caused by *Helminthosporium oryzae.* Ghewande (1987) observed that the aqueous leaf extracts of *Azadirachta indica, Lawsonia alba, Pongamia glabra* and *Tridax procumbens* at 2 per cent concentration were effective in reducing the rust infection of groundnut caused by *Puccinia arachidis* to an extent of 10.4 to 39.0 per cent over control and also increased yield ranging from 15.42 to 28.12 per cent over control.

Mathikumar (1989) reported that the non-edible seed oils of neem (*A. indica*) pinnai (*Calophyllum inophyllum*) Illupai (*Madhuca longifolia*) at 1.0 per cent and seed kernel extract of neem at 5.0 per cent concentration were also effective as fungicides in reducing the rust disease of groundnut though fungicides were more effective. Among the treatments tested baycor (0.2 per cent), calixin (0.05 per cent) and Illuppai oil (1.0 per cent) recorded least disease with a per cent disease index value (PDI) of 19.07, 32.95 and 34.44 respectively as against 61.85 in control treatment. The yield increase recorded from the plots treated with these preparations were 29.1, 22.0 and 19.7 per cent over control during rabi season of the year 1984. In the kharif season also similar results were obtained..

Thangamani Narayanasamy and Narayanasamy (1988) stated that neem oil and neem seed kernel extract spray (at 2 per cent) reduced sheath rot disease (*Sarocladium oryzae*) incidence to 10.7 and 8.6 per cent respectively as against 14.7 per cent in control.

Mariappan *et al.* (1988b) reported that application of neem oil (1.0 per cent) on 17 days after sowing in the nursery followed by six sprayings at weekly intervals in the mainfield or application of carbofuran in the nursery (2.5 kg a.i./ha) followed by neem oil spray in the mainfield as above reduced the rice sheath rot disease (*S. oryzae*) to 4.4 and 7.6 per cent respectively as against 19.3 per cent in control.

In the present study the efficacy of neem oil and neem seed kernel extracts were tested for their efficacy on the control of late blight disease of potato (*P. infestans*), Mycosphaerella leaf spot of cabbage (*M. brassicola*), blast disease of garlic (*B. allii*), phyllosticta leaf spot of ginger (*P. gingibari*).

The incidence of mycosphaerella leaf spot disease of cabbage was reduced significantly by both neem oil (0.4 per cent) and neem seed kernel extract (4.0 per cent) preparations and was comparable to that of the fungicides in efficacy. The yield obtained from plots treated with neem preparations were either similar to the yield obtained from plots treated with neem preparations were either similar to the yield obtained from plots treated with carbendazim and chlorothalonil and was even more than the yield obtained from treatments of Thiophanate methyl, copper oxychloride or mancozeb.

The blast disease of garlic has been reduced by both neem oil and neem seed kernel extract significantly and were comparable to the efficacy of fungicides. The fungicides Thiophanate methyl, chlorothalonil, carbendazim reduced the mean disease index to 0.96, 1.07 and 1.69 respectively compared to 5.73 in control. The neem oil reduced the mean disease index to 1.49 and thus proved to be as effective as fungicides. The above three fungicides recorded an yield of 3.65, 3.96 and 3.48 t/ha respectively while neem oil and neem seed kernel extract treated plots yielded 3.5 t/ha which are higher than the 1.98 t/ha from control plots. Thus neem preparations were found to be as effective as fungicides.

The phyllosticta leaf spot of ginger has been controlled effectively by neem preparations and were even more effective than most of the fungicides except chlorothalonil. The neem oil, neem seed kernel extract, chlorothalonil and carbendazim treatments recorded a mean disease index value of 1.82, 1.91, 1.00 and 1.98 respectively which are less than the other treatments. These treatments recorded a yield of 4.42, 4.43, 4.62 and 4.53 t/ha respectively which were higher than the other treatments and also that of the yield recorded from control plots of 2.72 t/ha.

The neem preparations were not relatively effective in reducing the potato late blight disease (*Phytophthora infestans*). However neem oil has been found to be effective against most of the diseases tested and were comparable to the existing recommended pesticides in its efficacy. Unlike the chemical pesticides the neem oil is not harmful and so far there are no report on its presence as residue in edible vegetables. The neem oil also is easily available in Tamil Nadu. Hence the possibility of using neem oil for the control of pests and diseases of vegetable crops deserve consideration.

REFERENCES

Chatur Vedi, R., D. Anupam, C. Dikshit and S.N. Dixit. 1987. *Adenocalynima alliaceum* a new source of natural fungitoxicant. *Trop. Agric.*, **64**(4): 318-382.

Duleep Kumar Samuel, L. 1990. Studies on chilli mosaic virus. M.Sc.(Ag.) Thesis. Tamil Nadu Agricultural University, Coimbatore, India. pp. 133.

Ghewande, M.P. 1987. Management of groundnut foliar diseases using plant extracts. *Indian Phytopath.*, **40**: 283.

Mariappan, V., S. Jayaraj and R.C. Saxena. 1988a. Effect of non edible seed oils on survival of *Nephotettix virescens* (Homoptera : Cicadellidae) and on transmission of Rice tungro virus. *J. Econ. Entomol.*, **29**(4): 395-411).

Mariappan, V., V. Narasimhan and R. Jeyarajan. 1988b. Effect of neem on rice disease. *In*: Final workshop of IRRI-ADB-EWC project on Botanical pest control in Rice Based Cropping systems held in December 12-18, 1988 at IRRI, Philippines (8).

Mathikumar, P. 1989. Studies on rust disease of groundnut. M.Sc. (Ag.) Thesis. Tamil nadu Agricultural University, Coimbatore, India. pp. 96.

Ravindra Babu. 1987. Studies on mungbean yellow mosaic virus disease. Ph.D. Thesis, Tamil nadu Agricultural University, Coimbatore, India. pp. 187.

Thangamani Narayanasamy and P. Narayanasamy. 1988. Role of neem products in the control of fungal disease of rice. *In*: National seminar on management of crop diseases with plant products/ biological agents. January, 11-12, 1988. pp. 6.

Neem for the Management of Crop Diseases, 1995
Editor : V. Mariappan
Associated Publishing Co., New Delhi, India
pp. 63-67

EFFECTIVENESS OF THE NEEM FOR THE MANAGEMENT OF VEGETABLE DISEASES

Roy, A.K.

University Department of Botany, Bhagalpur University
Bhagalpur-812 007, India

INTRODUCTION

The use of foliar extracts to control plant diseases was first attempted by Democritus in 470 BC (Sharville, 1960). Neem (*Azadirachta indica*), which is a common back yard and indigenous tree of India has established biocidal, medicinal, environmental purifier and wind breaker potentials. Its popularity is too old to be considered as national heritage whose virtues cited in the ancient literary classics. History witnesses that before the advent of synthetic pesticides, the farmers in India had protected their crops in field and storage from pests by using the different parts (particularly leaves and fruits) of neem. Such unique potentials of neem is due to presence of chemicals like Azadirachtin, meliantriol, salannin etc. of limnoid group which have antifungal, antibacterial, antiviral, anti-insect, repellant and toxicant properties. Recently its demand in European countries has increased in many folds. Several foreign industries are in practice to patent insecticides prepared from neem. The situation becomes alarming to India because the foreign scientists are paying more attention for the extraction of effective pesticides from neem. Extracts of some chlorophyllous plants like *Phaseolus vulgaris (Afifi, 1975), Rosa indica* (Dixit *et al.*, 1976), *Achyranthus aspera* (Chakrovarti and Pariya, 1977) showed inhibitory action on pathogenic fungi. Bhowmick & Vardhan (1982) have also reported the antimicrobial properties of some green plants including neem. In addition, there are several reports that neem products are used to control cucumber mosaic of virus and grassy stunt virus disease of rice. The foliar extract of neem has, thus, wide spectrum inhibitory effect on phytopathogens. However, specific study is still wanting. Keeping this in view the present investigation has been carried out to evaluate the effect of leaf extracts of neem on growth and pathogenic behaviour of fungi causing fruit rot diseases of gourd fruits (sponge, bottle & pointed).

MATERIALS & METHODS

A regular survey (monthly) of fields and local markets were conducted to record the fungal damage of sponge-gourd, bottle-gourd and pointed gourd fruits. The diseased samples were brought to the laboratory separately in polyethylene bags to avoid the intra-contamination. The organisms were isolated from the diseased fruits by following the usual method. Cultures were maintained on PDA/Asthana Hawker's medium A. Pathogenicity test of each organism was conducted (i) by placing inoculum directly on fruit surface (ii) by spraying spore suspension on injured fruit surfaces. Injuries were inflicted with sterilized needle and care was taken that injury should be superficial. Whenever more than one organisms was isolated from a particular spot, the pathogenicity of each organism was tested separately. Pathogenicity was confirmed where Koch's (1882) postulates were fully satisfied. The development of disease and extent of damage caused by each fungus were also studied. On the basis of extent of loss in field and market fungal organisms were selected for the present investigation. *Fusarium scirpi, Helminthosporium speciferum* were found to cause maximum loss of sponge gourd fruits, similary, *Curvularia lunata, F. equiseti* in case of pointed gourd and *F. semitectum* and *Botryodiplodia theobromae* in case of bottle gourd were selected for the present investigation.

In order to control the post harvest diseases of gourd fruits, leaf extracts of neem was used to record their effect on radial growth, mycelial out put and spore germination under laboratory condition. The effectiveness of leaf extracts was also evaluated on diseased fruits.

Young leaves of neem were collected from tree and leaf extract was prepared by following the method of Kumar & Sachan (1979). Leaves were thoroughly washed and aqueous extract (2:10 w/v) of leaves were prepared in glass homogenizer separately. The homogenized liquid was filtered through muslin cloth and centrifuged at 3000 rpm for 15 minutes. In order to assay antifungal activity of extract, observation on radial growth, mycelial dry wt. and spore germination were made.

For evaluating the effectiveness of leaf extract on controlling diseases the gourd fruits of same size and of same physiological age were given both pre-& post infection treatments (Sumbali & Mehrotra, 1982). In pre-infection treatments weighed fruits prior to inoculation were given a dip in the extract for 2-5 minutes and kept for 12 hrs in sterilized polyethylene bags and thereafter, fruits were inoculated with spore suspension of 7 days old culture by pin prick injury method in each case. In case of post infection treatment weighed fruits were first inoculated and incubated at 28 ± 2°C & for 12 hrs and then given a dip of leaf extract for the same period. Checks were always maintained by dipping fruits in sterilized distilled water. Observations based on average of three replicates for each case was recorded.

RESULTS & DISCUSSION

Effect of foliar extract on growth behaviour : *Fusarium semitectum & Botryodiplodia theobromae* isolated from bottle gourd rot, *F. equiseti, Curvularia lunata* from pointed gourd rot; and *F. scirpi* and *Helminthosporium speciferum* from sponge gourd rot were found to cause considerable loss of the respective fruits in field as well as in the market. These pathogens were selected to observe the effect of foliar extracts of neem on growth behaviour of these fungi. Results are depicted in Table-1. Colony diameter, mycelial dry wt. and % spore germination were taken into consideration. *F. equiseti* and *C. lunata* pathogens were found to be affected at the highest level in respect to growth behaviour and spore germination, however, *F. semitectum* and *F. scirpi* were affected at lower level (below 5 % of inhibition). *H. speciferum* and *B. theobromae* were affected moderately (more than 70%). A general view of the results also reveals that there was direct correlation of mycelial growth and sporulation. Dixit and his associates (1976) have reported that the leaf extract of some angiospermic plants exhibited strong toxicity for various pathogenic fungi.

Table 1. Effect of foliar extracts (2:10 w/v) of *Azadirachta indica* on growth behaviour and spore germination of gourd-rot fungi

Name of the fungi	Nature of treatment	Growth behaviour		Spore germinability	
		Colony diameter (cm.)	Mycelial dry wt. (mg.)	Germination (%)	Inhibition (%)
Fusarium scirpi	Control	7.0	450	74	
	Treated	5.8	415	44	40
F. equiseti	Control	4.1	312	74	
	Treated	0.3	28	5	93
F. semitectum	Control	7.5	400	81	
	Treated	7.0	360	42	48
Curvularia lunata	Control	3.5	218	60	
	Treated	0.1	15	2	96
Helminthosporium	Control	7.5	495	72	
speciferum	Treated	5.6	392	18	75
Botryodiplodia theobromae	Treated	5.2	175	20	71

Effect of foliar extract on control of fruit rots : Leaf extracts of neem were applied on pre- and post infected fruits by their respective fungi in order to evaluate the efficiency of the extract. Results are tabulated in Table-2. It is evident from the experimental findings that the post infection treatment showed comparatively more pronounced effect on control of fruit rots than pre-infection treatment. The treatment of foliar extracts after infection evaded the infection caused by *F. scirpi, H. speciferum* in case of sponge gourd and *F. equiseti* and *Curvularia lunata* in case of pointed gourd fruits. The pathogens i.e. *F. semitectum* and *B. theobromae* causing rots on bottle gourd fruits were also affected but their growth could not be checked completely. Similar response of leaf extract controlling fruit rot has also been observed by Pandey (1982) and Gourinath and Manoharchary (1984). Arya (1988) observed that 50% concentration of leaf extracts of *Eucalyptus occidentalis* and *Aegle marmelos* were effective to *phomopsis* fruit rot of grapes and guava.

The inhibitory effect of leaf extracts against the growth of the test fungi and their effect on fruit diseases may be attributed to the presence of active principles like alkaloids, steroids, terpenes and phenols (Singh, 1986, Prasad, 1986; Arya, 1988;

Kishore *et al.*, 1989 and Dubey & Dwivedi, 1991). Mahadevan (1982) has also reported that plant contains certain inhibitory substances "prohibitors" which are toxic to micro organisms. In case of foliar extracts of neem the Azadirachtin, Meliantriol, Salannin etc. compounds might have inhibitory effect on the growth behaviour and spore germination of fungi and subsequently active in controlling the rot of fruits.

Table 2. Effectiveneess of leaf extracts on control of fruit rots

Name of the pathogens	Sponge gourd		Bottle gourd		Pointed gourd	
	PR*	PO*	PR	PO	PR	PO
Fusarium scirpi	72.45	100				
Helminthosporium speciferum	60.55	100				
Fusarium semitectum			42.55	75.23		
Botryodiplodia theobromae			68.32	92.44		
Fusarium equiseti					82.43	100
Curvularia lunata					76.97	100

* Pre-infected gourd fruits
* Post-infected gourd fruits

In general the fruit rots of bottle gourd, sponge gourd and pointed gourd may be minimized by treating them with foliar extracts of neem particularly before marketing. Field spray has also been suggested to control the incidence of rot causing fungi. However, the present investigation has opened the scope for the researchers working on biochemistry to identify the antimycotic active compounds of neem and further this may also be exploited for the control of plant diseases. It is expected that the recent researches will enable to bring neem out from the temple court yards to the business front. Farmers of India can grow their own neem trees and produce simple and safe insecticides to manage the pests and earn substantial profit.

REFERENCES

Afifi, A.F. 1975. Effect of volatile substances from species of Labiatae on rhizospheric and phyllospheric fungi of *Phaseolus vulgaris. Phytopath. Z.*, **83**: 296-301.

Arya, A. 1988. Control of phomopsis fruit rots of grapes and guava. *Indian Phytopath.*, **42**(2): 214-219.

Bhowmick, B.N. and Vijay Vardhan. 1982. Antimycotic activity of leaf extracts of some medicinal plants on *Drechslera turcica* (Prac.) Subram. & Jain. *Biol. Bull. India*, **4**(1): 58-60.

Chakravarty, D.K. and S.N. Pariya. 1977. Inhibition of germination of phytopathogenic fungi in some Indian medicinal plant extracts. *Zeitschrift fur Pflanzenkrari - Khaiten* and *Pflanzenschutz*, **84**: 221-223.

Dixit, S.N., S.C. Tripathi and R.R. Upadhyaya. 1976. The antifungal substances of rosa flowers (*Rosa indica*). *Eco. Bot.*, **30**: 371-374.

Dubey, R.C. and R.S. Dwivedi. 1991. Fungitoxic properties of some plant extracts against vegetative growth and sclerotial viability of *Macrophomina phaseolina. Indian Phytopath.*, **44**(9): 411-41[illegible]

Gaurinath, A. and C. Manoharchary. 1984. Effect of angiospermic leaf extract on three pathogenic fung[illegible] *Indian Bot. Reptr.*, **3**(1): 95-97.

Kishore, N., S.N. Dixit and N.K. Dubey. 1989. Evaluation of leaves of some plant for their volatile fungitoxicity. *J. Indian Bot. Soc.*, **68**(2,3,4,): 167-168.

Koch, R. 1882. C.F. Riker, A.J. and R.S. Riker Introduction to Research on Plant Diseases (1936). John S. Surift Co. New York.

Kumar, R. and S.N. Sachan. 1979. Effect of some plant extracts on the conidial germination of *Curvularia pallescens. Indian Phytopath.,* **32**(3): 489.

Mahadevan, A. 1982. Biochemical Aspects of Plant Disease Resistance, Part-I. Today and Tomorrows Printers and Publishers, Delhi, pp. 110-164.

Pandey, K.N. 1982. Antifungal activity of some medicinal plants on stored seeds of *Eleusine coracana. Indian Phytopath.,* **35**(3): 499-501.

Prasad, A.K. 1986. Pathological investigation on some varieties of pointed gourd fruit common in Bihar. Ph.D. Thesis, Bhagalpur University, Bhagalpur.

Sharville, E.B. 1960. The nature and uses of modern fungicides. Burgecs Publishing Co. Minnesota, USA.

Singh, C.P. 1986. Pathological and Biochemical studies on Banana fruits (Vr. green) common in north Bihar Ph.D. Thesis, Bhagalpur University, Bhagalpur.

Neem for the Management of Crop Diseases, 1995
Editor : V. Mariappan
Associated Publishing Co., New Delhi, India
pp. 69-76.

EFFICACY OF BOTANICALS FOR THE MANAGEMENT OF BLIGHT DISEASE OF TWO VEGETABLE CROPS

Narasimhan, V., Vijayan, M. and Senthilnathan, V.

Tamil Nadu Agricultural University, Coimbatore-641 003, India

INTRODUCTION

Plant products have assumed special significance in the present day strategy of developing environmentally sound methods of plant disease control. Frequent and constant use of plant protection chemicals for disease control especially for vegetables have resulted in the problem of residual toxicity and pollution in the environment. But the use of botanicals avoid the risk of phytotoxicity, accumulation of harmful residues on plants and in soil. Since the first record of the effect of plant extract of *Acorus calamus L on Alternaria* sp. and *Helminthosporium* sp. (Mann and Burns, 1927), several reports have been published on various fungal, bacterial and viral pathogens of crop plants (Sbragia, 1975). With the increase in the awareness towards toxic hazards of pesticides to crops and environment, the importance of indigenous botanicals in plant disease control have been emphasized (Krishna Kumar and Nene, 1968).

Earlier reports on vegetables indicate that initial screening studies were conducted with several parts of the plants against *Alternia* species. The efficacy of leaf extracts of *Acacia loculate, Ficus religiosa, Amaranthus viridis, Melia azederach on A. tenuis, Nicotinia glutinosa on A. brassicola* (Shekhawat and Prasada, 1971), flower extract of *Lawsonia alba* on *A. alternata* (Pandey et al., 1981), root extract of *Achyranthus aspera on A. alternata* (Pandey et al., 1981) have been reported. However, detailed studies on the comparative effect with fungicide, stability of the compound and the active principle involved in fungitoxicity are lacking. Investigations carried out on the efficacy of various botanicals on two vegetables, viz., tomato and onion on the blight disease, their stability under different temperatures, effect during storage and the nature of antifungal componant are discussed in this paper.

Efficacy of plant extracts on the pathogens causing blight disease on tomato and onion : Preliminary screening of 44 plant extracts for testing their antifungal property against *Alternaria solani* (Ell. and Mart.) Jones and Grout causing blight disease of tomato was carried out by poisoned food technique and spore inhibition method. The results revealed that the bulb extract of garlic (*Allium sativum*), leaf extract of vilvam (*Aegle marmelos*) and flower extract of red periwinkle (*Catheranthus *roseus*) at 10 percent concentration prepared in water were most effective in inhibiting the spore germination and mycelial growth of the pathogen (Table-1) Among the eight plant species screened against *Alternaria tenuissima* (Kunze) Wilt Shire causing blight disease of onion, the leaf extracts of *Aegle marmelos* and *Prosophis juliflora* were effective in inhibiting the spore germination and mycelial growth (Table-2).

Table 1. Effect of plant extracts/products on the spore germination and mycelial growth of *Alternaria solani*

Sl. No.	Common name	Botanical name	Spore germination*		Mycelial growth	
			Ger.(%)	Inh.(%)	Grw.(mm)	Inh.(%)
1.	Onion bulb	*Allium sativum*	17.8	90.3	12.7	85.9
2.	Neem seed	*Azadirachta indica*	24.8	81.6	21.7	75.9
3.	Neem cake	- do -	42.5	52.3	64.0	28.9
4.	Neem leaf	- do -	22.4	84.8	21.3	76.3
5.	Vilvam leaf	*Aegle marmelos*	19.3	88.6	14.7	83.7
6.	Betel leaf	*Piper betle*	20.2	87.5	19.0	78.9
7.	Red periwinkle	*Catheranthus roseus*	18.8	89.1	15.3	83.0
8.	Allitin	Synthetic product (based on the active principle in *A. sativum*)	14.7	93.3	10.0	88.9
9.	Control	Sterile water	77.6	–	90.0	–
	CD (P = 0.05)		3.7		13.3	

(Mean of three replications) * Data after angular transformation

Table 2. Effect of plant extracts/products on the spore germination and mycelial growth of *Alternaria tenuissima*

Sl. No.	Common name	Botanical name	Spore germination* (%)	Mycelial growth (%)
Bulb extract:				
1.	Garlic	*Allium sativum* L.	52.7	52.3
Leaf extract:				
2.	Pome granate	*Punica granatum* L.	50.2	48.3
3.	Gundumani	*Adenanthera pavonia* L.	48.0	47.7
4.	Neem	*Azadirachta indica* A. Juss	46.1	43.3
5.	Betel vine	*Piper betle* L.	50.4	49.7
6.	Vilvam	*Aegle marmelos* Corr.	22.1	10.3
7.	Prosopis	*Prosophis juliflora* L.	25.5	14.7
8.	Red gulmohar	*Delonix regia* Ref.	46.0	40.0
9.	Control	Sterile water	80.6	78.7
	CD (P = 0.05)		2.1	3.2

(Mean of three replication) * Data after angular transformation

Efficacy of vegetable oils on the blight pathogens : Vegetable oils were tested against the two foliar pathogens. Among the four seed oils tested against *A. solani*

neem seed oil was the most effective treatment with 57 per cent inhibition of spore germination and 35.8 per cent of mycelial growth (Table-3).

Table 3. Effect of seed oils on spore germination and mycelial growth of *A. solani*

Common name	Botanical name	Spore germination*		Mycelial growth	
		Ger. (%)	Inh. (%)	Grow. (mm)	Inh. (%)
Neem	*Azadirachta indica*	40.9	57.0	56.7	35.8
Illuppai	*Madhuca longifolia* var. *latifolia*	45.0	49.6	67.3	23.8
Pinnai	*Calophyllum inophyllum*	45.5	48.8	66.7	24.5
Castor	*Ricinus communis*	41.7	55.6	72.7	17.7
Control	Sterile water	99.5	–	88.3	–
	CD (P = 0.05)	5.1		12.8	

Mean of three replication * Data after angular transformation

Seed oils were also tested against *A. tenuissima* and among them Iluppai oil could cause maximum inhibition of spore germination (50.1%) and mycelial growth (46.5%). (Table-4).

Table 4. Effect of seed oils on spore germination and mycelial growth of *A. tenuissima*

Common name	Botanical name	Spore germination*		Mycelial growth	
		Ger. (%)	Inh. (%)	Grow. (mm)	Inh. (%)
Neem	*Azadirachta indica*	58.2	25.6	59.8	24.5
Pinnai	*Calophyllum inophyllum*	60.3	22.3	63.8	19.4
Illuppai	*Madhuca longifolia* var. *latifolia*	44.1	50.1	42.4	46.5
Control	Sterile water	80.2	–	79.2	–
	CD (P = 0.05)	1.6		1.7	

Mean of three replication * Data after angular transformation

Comparative efficacy of selected plant extracts/products and fungicides: Selected plant extracts (extracted in water 10%) were evaluated against *A. solani* and *A. tenuissima* in comparison with the standard effective fungicide (Table-5). All the plant extracts tested were effective against *A. solani* and *A. tenuissima* in inhibiting spore germination and mycelial growth, but were only next to the standard fungicide Dithane M. 45, Cuman L. and Bavistin.

Stability of the antifungal activity of the extracts at different temperatures: The antifungal activity of selected products were tested for their stability by heating the extracts in waterbath at different temperatures viz., 30.2 (room temp.), 50.0, 70.0, and 90.0 c. Heating the plant extracts at high temperature (90.0 c) reduced the antifungal activity against *A. solani.* Among them *A. sativum* was the most effective and retained their activity upto 70.0 c. In a similar test conducted with *A. tenuissima* also the antifungal activity significantly reduced when heated at high temperature (90.0 c), while at lower temperatures they retained their activity.

Table 5. Effect of different plant extracts (10%) in comparison with fungicides

Plant extract	Spore germination* (%)	Mycelial growth (mm)
A. Against *A. solani*		
A. sativum bulb	25.6	14.0
A. marmelos leaf	28.5	19.0
P. betle leaf	28.6	74.0
A. indica leaf	45.8	72.0
A. indica seed	45.6	75.3
C. roseus flower	28.2	21.0
Dithane M-45 0.2%	12.0	10.3
CD (P = 0.05)	2.0	3.9
B. Against *A. tenuissima*		
Aegle marmelos leaf	23.8	13.5
Prosophis juliflora leaf	29.5	20.0
Cuman L. 0.2%	1.8	0.0
Bavistin 0.1%	18.6	9.8
Control	78.8	84.1
CD (P = 0.05)	2.8	2.0

Mean of three replication   * Data after angular transformation

The stability of the antifungal activity during storage when tested 0° to 20° revealed that *A. sativum* bulb extract retained their inhibitory effect on spore germination and mycelial growth of *A. solani* from 0 to 20°C upto 10 days. *C. roseus* flower and *A. mermelos* leaf extracts retained their effect upto 5 days. All the other plant extracts retained their effect upto one day only against both the pathogens.

Persistence of antifungal activity: The water extracts from *A. sativum* bulb and *A. marmelos* leaf persisted on the foliage upto 5 days after spray on the foliage (more than 60% inhibition) when tested with *A. solani* and this was on par with the standard fungicide, Dithane-M 45 (0.2%) (Table-6).

Table 6. Persistence of antifungal effect of plant extracts (10%) on tomato leaves against *A. solani* as determined by spore germination and mycelial growth

Name of the plant	Days after treatment with the plant extracts					
	Spore germination (%)			Mycelial growth (mm)		
	0	5	10	0	5	10
A. sativum bulb	4.1	20.2	43.2	12.7	48.3	57.3
A. marmelos leaf	4.1	28.6	52.4	19.3	54.7	76.3
C. roseus flower	23.6	46.7	62.3	17.7	53.7	78.3
Dithane M-45 00.2%	4.1	19.5	41.3	10.7	31.7	52.7
Control	80.2	80.8	79.3	75.0	85.3	79.3
CD (P = 0.05) Treatment			2.0			2.0
Days			2.2			2.2

In respect of the onion pathogen, *A. tenuissima,* the antifungal activity of *A. marmelos* leaf extract persisted on the foliage upto 5 days (Table-7).

Table 7. Persistence of antifungal effect of plant extracts (10%) on tomato leaves against *A. tenuissima* as determined by spore germination and mycelial growth

Name of the plant	Days after treatment with the plant extracts					
	Spore germination (%)			Mycelial growth (mm)		
	0	5	10	0	5	10
A. marmelos leaf	19.0	3.5	72.0	8.0	30.0	73.0
Prosophis juliflora leaf	24.0	41.0	77.0	13.0	38.0	76.0
Cuman L 0.2%	2.0	23.0	52.0	0.0	12.0	59.0
Control	82.0	82.0	81.0	79.0	78.0	78.0
CD (P = 0.05) Treatment Days		0.7			0.7	
			0.8			0.8

Management of seedborne infection of *A. solani* on onion with botanicals: Among the different plant extracts evaluated against *A. solani* on tomato seeds by seed soaking in plant extract solution, the water extracts (10%) from *A. sativum* bulb and *C. roseus* flower were effective in increasing the seed germination and seedling vigour than the standard fungicide, thiram @ 2g/kg of seed (Table-8).

Table 8. Effect of plant extracts *A. solani* treated seeds on seed germination and seedling vigour of tomato seed

Treatment	Germination (%)	Seedling vigour index (No)
A. sativum bulb + A. solani	52.8	615
A. marmelos leaf + A. solani	50.8	582
C. roseus flower + A. solani	52.8	672
Thiram + A. solani	54.8	567
A. solani alone	35.2	57
Control	50.8	426
CD (P = 0.05)	6.6	58

Management of foliar infection of *A. solani* on tomato and *A. tenuissima* on onion under *in vivo* conditions: The effectiveness of plant extracts (10%) against foliar infection were tested under pot culture conditions on blight diseases intensity on tomato and onion by pre and post inoculation spray treatments and compared with the standard fungicide (Table-9).

Pre-inoculation spray with *A. sativum* bulb and *A. marmelos* leaf extracts afforded greater protection against tomato blight disease intensity than *C. roseus* flower extract, but was comparatively less effective than *A. marmelos* leaf extract. Post inoculation spray was comparatively less effective in reducing the disease intensity. However, the effect of these plant extracts were next to the fungicide, dithane-M45.

Table 9. Effect of pre and post inoculation spray with plant extracts (10%) on early blight intensity on tomato

Plant extracts	Pre-inoc. spray (Mean PDI)	Post-inoc. spray (Mean PDI)
A. sativum bulb	16.7	17.9
A. marmelos leaf	18.7	15.8
C. roseus flower	20.6	26.7
Dithane-M 45 0.2%	13.0	16.2
Control	55.9	63.4
CD (P = 0.05)	1.0	1.8

In a similar study with onion blight disease, *A. marmelos* leaf extract was the most effective treatment next to that of the fungicide, cuman-L (Table-10).

Table 10. Effect of pre and post inoculation spray with plant extracts (10%) on early blight intensity on onion

Plant extracts	Pre-inoc. spray (Mean PDI)	Post-inoc. spray (Mean PDI)
A. marmelos leaf	32.7	29.1
Prosophis juliflora leaf	40.7	37.4
Cuman-L 0.2%	22.4	17.8
Control	51.2	51.5
CD (P = 0.05)	0.9	0.9

Field evaluation of the plant extracts with botanicals : Field experiments conducted against the blight disease on tomato with selected plant extracts revealed that *A. marmelos* leaf extract (10%) as foliar spray once at the initial symptom appearence could effectively check the spread of the disease and the effect was next to that of the standard fungicide, Cuman L (0.2%). (Table-11).

Table 11. Efficacy of selected plant extracts (10.0%) on the intensity leaf blight of tomato caused by *A. solani* under field condition

Plant extract	Mean blight intensity (PDI)
A. marmelos leaf	28.8
Prosophis leaf	32.4
Cuman-L (0.2%)	18.3
Control	33.2
CD (P = 0.05)	1.0

PDI : Per Cent Disease Index

Detection of the anti fungal componant of the plant extracts: The crude plant extracts from the most effective plant species were separated into protein and non protein parts by the method of Stumpf, 1955 and evaluated for their efficacy against the blight pathogens under *invitro* conditions. In all the plant extracts tested the protein part of the extract had the maximum inhibitory effect on spore germination and mycelial growth of *A. solani* and *A. tenuissima* (Table-12).

Table 12. Efficacy of protein and non protein components of the plant extracts on spore germination and mycelial growth of *A. solani* and *A. tenuissima*

Pathogen	Plant extract	Spore germination (%)		Mycelial growth (mm)	
		Protein part	Non protein part	Protein part	Non protein part
A. solani	*A. sativum* bulb	13.9	61.6	11.7	75.7
	A. marmelos leaf	16.9	56.4	13.3	63.3
	C. roseus flower	18.9	62.5	61.7	72.7
	Control (Water)	83.2	82.5	90.0	90.0
	CD (P = 0.05)	3.3		2.8	
A. tenuissima	*A. marmelos* leaf	14.2	61.4	3.0	49.5
	Prosophis juliflora leaf	21.8	73.1	10.5	67.8
	Control (Water)	83.0	83.0	78.8	78.8
	CD (P = 0.05)	4.0		2.4	

The studies have indicated that the active principle responsible for the antifungal activity in the extracts are proteinaceous in nature. Earlier studies done by other workers reveal that extracts from garlic possessed antifungal activity against other plant pathogenic fungi like *Pyricularia oryzae, Colletotrichum capsici* and *Glomerella cingulata* (Murthy and Amonker, 1974) and the active principle in this extract was reported to be due to the presence of "Allicin" which has antimicrobial activity (Cavellito and Bailey, 1944). In *A. marmelos* leaf the presence of "Fumocoumarin" compound might be responsible for the inhibitory effect (Manjunath, 1948).

CONCLUSION

The present studies and available reports have shown that botanicals could be successfully used for effective management of vegetable diseases. Though the antifungal effect was next to that of the standard fungicides tested, they are environment friendly and cheap sources for the grower. Avoiding the hazards of toxic chemicals especially for the vegetables crops is essential and the botanicals have a potential alternate source for the cultivators.

REFERENCES

Ahmed, S.K. and J.P. Agnihotri. 1977. Antifungal properties of some plant extracts. *Indian J. Mycol. Pl. Pathol.*, **2**: 180-181.

Cavellito, C.J. and J.H. Bailey. 1944. Allicin, the antibacterial principle of *Allium sativum*. I. Isolation, physical properties and antibacterial action. *J. Amer. Chem. Soc.,* **66**: 1950-1951.

Chakravarty, D.K. and S.N. Pariya. 1977. Inhibition of germination of phytopathogenic fungi in some indian medicinal plants. *Z. Pflanzenkarankh, Pflanzenschutz.,* **84**: 221-223.

Krishnakumar and Y.L. Nene. 1968. Antifungal properties of *Cleome isocandra* L. extracts. *Indian Phytopath.,* **21**: 445-446.

Mulkadam, D.S., L.D. Balkande and G.V. Umalkar. 1976. Antifungal activities in deproteinized leaf extracts of weeds and non weeds. *Indian J. Microbiol.,* **16**: 78-79.

Murthy, N.B.K. and S.V. Amonker. 1974. Effect of natural insecticide from garlic (Allium sativum L.) and its synthetic form (Diallyl-disulphide) on plant pathogenic fungi. *Indian J. Expt. Biol.,* **12**: 208-209.

Manjunath, B.L. 1948. *The wealth of India.* Vol. I. Publications information directorate, CSIR, New Delhi. 253 p.

Mann and Burns. 1927. Pharmocology and chemotherapy of plant products. In: *"The Indian Materia Medica".* School of tropical medicine, Calcutta, pp. 63-67.

Pandey, D.K., R.N. Tripathi and N.N. Tripathi. 1981. Antifungal activity in some seed extracts, *Environment India.* **4**: 83-85.

Sbragia, R.J. 1975. Chemical control of plant diseases: An exciting future. *Ann. Rev. Phytopath.,* **13**: 257-267.

Shekhawat, P.S. and R. Prasada. 1971. Antifungal properties of some plant extracts. II. Growth. Inhibition studies. *Sci. & Cul.,* **37**: 40-41.

Neem for the Management of Crop Diseases, 1995
Editor : V. Mariappan
Associated Publishing Co., New Delhi, India
pp. 77-80.

MANAGEMENT OF BLACK SCURF OF POTATO CAUSED BY RHIZOCTONIA SOLANI BY USING NEEM CAKE

S. Singh, Rama, Singh, Jagpal, Jindal, Anuradha and Kapur, S.P.

Department of Plant Pathology, Punjab Agricultural University, Ludhiana

In modern age of intensive croping, the disease problems become one of the limitations for the crop production. The occurrence of black scurf of potato caused by *Rhizoctonia solani* is very common and sometimes pathogen damaged the crop upto great extent by causing rotting of the sprouts or cankers which lead yellowing and ultimately death of the plants (Blaszczak *et al.*, 1977). Plant disease control using toxic chemicals are not preferred now a days due to several problems such as environmental pollution and toxicity in food materials etc. The use of chemicals is also costly affairs for certain plant diseases. Therefore, some alternative methods in place of chemical control is being preferred for the management of diseases. Antagonistic microbial population (bio-fungicide) or some plant bio-cides (plant extract product) are used to minimize plant pathogens, hence, the reduction of disease occurrence. The products of neem (*Azadirachta indica* A. Juss) in various forms are known to have insecticidal and nematicidal properties. It also minimised the damage caused by certain fungi, bacteria etc. (Parmar and Singh, 1993; Schmutterer, 1990). The present investigation has been taken up with prime objective to observe the effects of neem cake on the growth of *Rhizoctonia solani* as well as on the development of black scurf on potato tubers.

MATERIALS AND METHODS

The *Rhizoctonia solani* was isolated from black scurf samples collected from experimental area of Punjab Agricultural University, Ludhiana. The cultures were purified and maintained on potato dextrose agar (PDA) medium for the experiments. The PDA medium was poisoned with neem cake in three concentrations i.e. 0.25, 0.50 and 1.0 per cent (W/V). The 20 ml agar medium was poured in sterlized petri dishes. After solidifying the medium small circular discs of *Rhizoctonia solani* from 6 days old culture was inoculated in the centre of each petri plate. The colony growth of pathogen was measured after 96 hours. The observations regarding width of mycelial rings, pattern of ring formation and sclerotial intensity were observed after 15 days of incubation. The inhibition of mycelial growth, colony development and sclerotial characters were also observed on potato dextrose broth poisoned by neem cake. The four replicates were kept for each treatment.

The effect of neem cake on the development of black scurf on potato tubers were observed in field for two seasons i.e. 1989-1990 and 1990-1991. The infected seed tubers were dipped in 1 per cent neem cake suspension for 10 minutes. Treated tubers were dried in shade and sowing was done after 24 hours. Finally the disease intensity on potato tubers was recorded following the method of Mayee and Datar (1986).

RESULTS AND DISCUSSION

Effect of neem cake on colony growth of *Rhizoctonia solani* on agar medium: The colonies growth of *Rhizoctonia solani* on PDA medium containing various concentrations of neem cake are presented in Table 1. The 89.6 mm colony diameter of pathogen was recorded after 96 hours of incubation in absence of neem cake. However, only 22 and 55 mm mycelial growth was observed on the medium containing 0.50 and 0.25 per cent neem cake respectively. The media with one per cent neem cake did not allow the growth of pathogen.

Table 1. Effect of neem cake on mycelial growth of *Rhizoctonia solani* on potato dextrose medium

Concentration of neem cake (%)	Colony diameter* (mm)	Inhibition (%)
0.25	55	38.7
0.50	22	75.6
1.0	0	100.00
0.0 (Control)	89.6	0

* Average colony growth of *Rhizoctonia solani* of four replicates after 96 hours of incrubation

The colony characters in relation to aerial mycelium, mycelial branching, width of mycelial ring and pattern of ring formation were also deviated significantly on the medium containing various concentration of neem cake as compared to control (Table 2). With increased concentration of neem cake in the medium, the abundance of fluffy aerial mycelium and mycelial branching of pathogen was observed, however, the width of mycelial rings decreased significantly on agar plates. Sclerotial intensity also decreased significantly with increase the neem cake concentration in the medium. There was only 15 per cent sclerotia observed at 0.5 per cent concentration of neem cake as compared to 60 per cent sclerotial intensity in absence of neem cake.

Effect of neem cake on growth and colony chatacter of *Rhizoctonia solani* on broth medium: The potato dextrose broth medium containing 0.25, 0.5 and 1.0 per cent neem cake was inoculated with *Rhizoctonia solani* causing black scurt of potato and mycelial growth was recorded after 15 days of incubation (Table 3). The 260 mg mycelial dry weight per flask of pathogen was observed in absence of neem cake as compared to 125 and 65 mg at 0.25 and 0.5 per cent concentration of neem cake respectively. Very little mycelial growth without sclerotia was observed on the broth medium containing 1 per cent neem cake. The colony characters also deviated significantly on the medium containing neem cake as compared to control. The colony on broth medium containing 0.5 per cent neem cake was smooth to wrinkled with moderate aerial mycelium as compared to smooth colony with suppressed growth in absence of neem cake. Similarly, 50 per cent inhibition in sclerotia formation was observed in broth medium containing 0.5 per cent neem cake. Brown to black coloured mycelial exudates were also observed on the mycelial mat grown on medium containing neem cake as compared to no exudates on mycelial mats grown on the medium without neem cake.

Table 2. Effect of neem cake on morphological/sclerotial characters of *Rhizoctonia solani* on potato dextrose agar medium

Colony/Sclerotial characters	Conc. of neem cake in medium (%)		
	0	0.25	0.5
Aerial mycelium	+	++	+++
Mycelial branching	+	++	+++
Width of mycelial rings (mm)	14.0	6	2
Pattern of ring formation	Circular	Circular	Wavy with zonation
Sclerotial intensity (%)	60.0	35.0	15.0
Inhibition of sclerotia formation (%)	0.0	41.6	75.0

+ Scanty, ++ Moderate, +++ Abundant

Table 3. Effect of neem cake on mycelial growth of *Rhizoctonia solani* in potato dextrose broth

Concentration of neem cake (%)	Mycelial dry* wt. (mg)	Inhibition (%)
0.25	125	52.0
0.50	65	75.0
1.0	20	92.3
0.0	260	0.0

* Observations taken after 15 days of incubation and are average of 4 replicates.

Effect of neem cake on development of Black scurf: The effect of seed tubers treatment in 1 per cent neem cake was observed for two seasons continuously (Table 4). There was only 8 per cent disease intensity observed in treated plots as compared to 49 per cent disease intensity in control plots during 1989-90 season. Similarly, during 1990-91, the treatment with neem cake gave only 10.1 per cent disease intensity on potato tubers as compared to 51.3 per cent black scurf in untreated plots.

The neem is indigenous to India and is cultivated throughout the country on account of its medicinal properties. Various plant parts/products are known to be used as medicine, insecticides, nematicides and fungicides (Parmar and Singh, 1993). In present study, the treatment of neem cake as seed treatment was found excellent to minimise black scurf of potato caused by *Rhizoctonia solani*. Its use as seed treatment or soil amendments was reported to have significant impact on soil borne plant pathogens. The soil application of neem cake was found very effective in management of several plant diseases caused by nematodes (Tyagi and Alam, 1993; Gupta *et al.,* 1993). Singh and Pandey (1967) reported significant reduction of *Pythium aphanidermatum* due to soil amendments with neem cake at higher temperature. The inoculum density of *Rhizoctonia solani* causing sheath blight of rice gradually declined after 6 months in the soil amended with neem cake. The pathogen could not be isolated after 11 months of incubation period (Kannaiyan and Prasad, 1981). Insecticidal and nematicidal properties of neem and its products have been well-documented (Rice, 1987; Saxena, 1987). However, delivery system and quantifying its effect on fungal pathogen needs throrough investigation.

Table 4. Effect of neem cake on development of black scurf due to *Rhizoctonia solani* on potato tubers

Year	Treatment*	Disease intensity (%)	Disease inhibition (%)
1989-90	Treated	8.0	83.7
	Untreated	49.0	0
1990-91	Treated	10.1	80.3
	Untreated	51.3	0

* Infected seed tubers were dipped in 1 per cent neem cake for 10 minutes.

REFERENCES

Blaszczak, W., S. Weber and M. Manka. 1977. The effect of temperature upon the pathogenecity and chemical control of *Rhizoctonia solani* K. on potato tubers (*Solanum tuberosum* L.) *Acta Agrobotanica,* **30**: 159-165.

Gupta, P., S. Sebastian, L.R. Sobita and S.M. Mishra. 1993. Use of neem cake in control of lesion nematode, cyst nematode and root knot nematode. *In*: International Neem Conference, Bangalore. pp. 37 (Abst.).

Kannaiyan, S. and N.N. Prasad. 1981. Effect of certain oil cakes on saprophytic activity of *Rhizoctonia solani* Kuhn in soil. *Indian. J. Microbiol.,* **21**: 79-80.

Mayee, C.D. and V.V. Datar. 1986. Phytopathometry. Technical Bulletin-1 (special Bulletin-3). Marathwada Agricultural University, Parbhani, pp. 146.

Parmar, B.S. and R.P. Singh. 1993. Neem in agriculture. IARI, New Delhi, pp. 85.

Rice, M.J. 1987. Neem Research in Australia. 11th Internation Congress of Plant Protection, Philippines, pp. 45-46. (Abst.).

Saxena, R.C. 1987. A decay of Neem research against rice insects pests in the Philippines, 11th ICPP, Philippines, pp. 46. (Abst.).

Schmutterer, H. 1990. Properties and potential of natural pesticides from the neem tree, *Azadirachta indica. Ann. Rev. Entomol.,* **35**: 271-297.

Singh, R.S. and K.R. Pandey. 1967. Population dynamics of *Pythium aphanidermatum* in oil cake amended soil. *Canadian J. Microbiol.,* **13**: 601-610.

Tyagi, S.A. and M.M. Alam. 1993. Efficacy of neem cake against soil nematodes and fungi. *In*: International Neem Conference, Bangalore. pp. 39-40. (Abst.).

Neem for the Management of Crop Diseases, 1995
Editor : V. Mariappan
Associated Publishing Co., New Delhi, India
pp. 81-84.

THE MANAGEMENT OF SOIL BORNE PATHOGENS OF GINGER WITH NEEM PRODUCTS

Dohroo, N.P. and Khan, M.L.

Dr. Y.S. Parmar University of Horticulture & Forestry
P.O. Nauni-Solan-173 230 (H.P.)

Ginger (*Zingiber officinale* Roscoe) is one of the important spice crops of India. The crop is affected by a number of diseases caused by various soil borne pathogens viz., species of *Pythium, Fusarium* and root knot (*Meloidogyne*) and root lesion (*Pratylenchus*) nematodes. Neem (*Azadirachta indica* A. Juss) products are well known for its pesticide potentialities having fungicidal and nematicidal activities against phytophagous and mycophagous nematodes. The principal active ingredient Azadirachtin and other limonoids such as meliantriol and salanin are quite effective against diseases of diverse nature. Besides, the decomposition of neem products in soil changes the rhizosphere microbiota thus adversely affect biocycle of pathogen (Khan *et al.,* 1974; Goswami Bhattacharya, 1989). The use of neem products in soil against soil borne pathogens of ginger manages different diseases of ginger by inactivation or lysis of spores, chlamydospores or hyphae either directly or due to production or release of antibiotics, competition for nutrients or parasitism.

Fungitoxic Properties: Several workers have reported the fungicidal property of neem cake and oil. It is a well known fact that the decomposition of neem products helps in alteration of physical, chemical and biological conditions of the soil thus reducing the inoculum potential of soil borne pathogens.

The neem cake amendment (500g/m^2) was found to suppress the population of *Fusarium oxysporum* and *Pythium ultimum* in soil (Table 1). *F. oxysporum* is a vascular pathogen causing yellows and *P. ultimum* causes rhizome rot of ginger.

Table 1. Effect of neem cake amendment on the population of *Fusarium oxysporum* and *Pythium ultimum* in soil

Treatment	CFU/ml of fungi	
	Fusarium oxysporum	*Pythium ultimum*
Neem cake	1×10^4	2×10^4
Check	4×10^4	3×10^4

Sharma (1989) studied the fungicidal properties of eight soil amendments against *F.o.* f.sp. *zingiberi* (Table 2).

Table 2. *In vitro* evaluation of soil amendments against *Fusarium oxysporum* f. sp. *zingiberi*

Treatment	Per cent growth inhibition
Cotton seed cake	24.26
Mustard cake	21.69
Sunflower cake	14.46
Neem cake	42.77
Sesamum cake	38.72
Wood saw dust	39.01
Farm yard manure	13.11
Dry pine needles	41.66

Neem cake was found best and inhibited the growth of the fungus to 42.7 per cent over check followed by pine needles. In a field experiment on the control of rhizome rot of ginger, the treatment of rhizomes in a combination of mancozeb and carbendazim and application of neem cake powder in soil managed the pre-emergence rot and yellows of ginger with simultaneous increase in yield (Table 3).

Table 3. Effect of seed treatments and neem soil application on the incidence of rhizome rot and yield of ginger

Treatment	Conc. (%)	Disease incidence (%)		Yield
		Pre-emergence	Yellows	(kg/plot)
Hexacap	0.25	5.5	4.0	7.0
Ridomil MZ	0.40	8.5	2.0	5.5
Blitox-50	0.30	9.0	6.5	5.0
Dithane M-45 +	0.25			
Bavistin +	0.10	4.0	3.5	7.6
Neem cake powder 500g/m^2				
Dithane M-45	0.25	17.4	8.5	4.2
Bavistin	0.10	2.5	3.0	7.5
Control	—	31.5	9.0	3.3
			CD(0.05%)	1.4

Plot size = 3m^2

The neem oil has been reported to contain some substances which have fungicidal properties (Vir and Sharma, 1984).

Biocontrol Properties: The use of neem products in soil is a type of cultural procedure which promotes natural biocontrol. Cultural treatments resulting in change in the environment of soil borne pathogens are the longest established methods of disease control. Various green or partially degraded crop residues and manures have been shown to control a wide range of pathogens including *Fusarium* spp.,

Rhizoctonia solani, Verticillium albo-atrum, V. dahliae and *Sclerotium* spp. (Jarvis and Thorpe, 1981; Lumsden *et al.* 1983). The three main direct mechanisms are parasitism or predation of one organism by another; antibiosis, where antagonists secrete metabiolities harmful to plant pathogens, and competition, where demand exceeds immediate supply of nutrients or space. The field experiments were conducted during 1991 and 1992 to see the effect of organic amendments on the incidence of the disease (Dohroo, 1993). The results revealed that the incidence of rhizome rot was minimum in pine needles and neem cake powder soil amendments (each 4 kg/3m^2). These amendments also increased the yield of ginger (Table 4).

Table 4. Effect of organic amendments on the incidence of rhizome rot of ginger

Treatment	Disease incidence (%)			Yield (kg/3m^2)		
	1991	1992	Mean	1991	1992	Mean
Neem cake powder	4.0	1.0	2.5	5.75	6.23	6.0
Wood saw dust	8.2	10.0	9.1	2.70	3.45	3.0
Pine needles	2.4	2.0	2.2	6.00	4.10	5.0
Oak leaves	8.5	6.0	7.2	3.00	5.73	4.4
Control	10.0	12.0	11.0	2.0	1.86	1.9

The rhizosphere mycoflora studies revealed that the population (CFU/ml) of *Trichoderma* and *Gliocladium* was maximum when soil was amended with neem cake powder (Dohroo, 1993).

Nematoxic Properties: The neem products have been shown to possess nematicidal properties. It is known that neem cake releases some nematoxic substances during their degradation in the soil. The various compounds released during decomposition, are ammonia, fatty acids, formaldehyde and phenolic compounds. The studies conducted by Khan (1989) on the integrated control of root-knot nematode in ginger revealed that application of neem cake when applied before planting was quite effective in the control of root-knot nematode and increased yield of ginger (Table 5).

Table 5. Effect of integrated control methods on ginger yield and nematode population

Treatment	Yield (kg/ha)	Final *M. incognita* juvenile/250 ml soil
Neem cake (Before planting)	6780	46.50
Carbofuran (Before planting)	5125	35.00
Neem cake (After 45 days of sowing)	4188	53.00
Carbofuran (After 45 daysof sowing)	7533	51.25
Check	4968	59.00

It was found that the soil application of neem cake @ 2.5 t/ha at the time of planting increased the yield, reduced the initial nematode population build-up critical for causing infection and simultaneously decreased the gall index.

Table 6. Effect of integrated control mathods on ginger yield and nematode population

Treatment	Yield kg/ha	Nematode Population		Gall index
		Initial	Final	
Neem cake @ 2.5 t/ha at planting	12,875	180	250	2.0
Carbofuran @ 3kg/ha at planting	9,500	200	210	2.5
Neem cake (45 DAP)	10,000	280	240	3.2
Carbofuran (45 DAP)	8,250	325	130	2.8
Control	6,675	400	450	3.4

FUTURE OUTLOOK

Soil borne plant pathogens play a major role in the root diseases of complex nature causing damping-oof, root rot, collar rot and wilt on many important field crops. These pathogens produce a variety of propagules such as oospores, sclerotia, chlamydospores, etc. which can survive in the soil for a considerable period of time. The role of neem products in the control of soil borne pathogens will involve their effect on the relative competitive advantage of pathogen versus antagonists. The studies are warranted on the effect of neem products on the population biology and ecology of antagonists in the microbiome of use. Not`least, the oil content of the neem products after the addition of auxilliary material is of utmost value in the management of soil borne plant pathogens. Keeping in view, the selective mode of action of neem products, these could best be used in Integrated Pest Management.

REFERENCES

Anon. 1990. Annual Report of the Department of Mycology and Plant Pathology for the year 1990-1991, UHF, Solan (HP): 94-95.

Dohroo, N.P. 1993. Final Report of Multilocational Project on Rhizome Rot of ginger (ICAR Cess Fund Scheme), Department of Plant Pathology, UHF, Solan. pp. 38.

Goswami, B.K. and D. Bhattacharya. 1989. Efficacy of neem and groundnut oil cakes as influenced by microorganisms. Changes in rhizosphere mycoflora during the decomposition of oil cakes in sterilized and unsterilized soil. *Indian. J. Nematol.* **19**(1): 72-74.

Jarvis, W.R. and H.H. Thorpe. 1981. Control of fusarium foot and root rot of tomato by soil amendment with lettuce residues. *Can. J. Pl. Path.*, **3**: 159-162.

Khan, M.L. 1989. Effect of integrated control methods on ginger yield and nematode population. Ann. Rept. Department of Entomology for the year 1989 and 1991, UHF, Solan 11 and 33-34.

Khan, M.W., M.M. Alam, A.M. Khan and S.K. Saxena. 1974. Effect of water soluble fractions of oil-cakes and bitter principles of neem on some fungi and nematodes. *Acta Bot. Indica*, **2**: 120-128.

Lumsden, R.D., J.A. Lewis and G.C. Ppapavizas. 1983. Effect of organic matter on soil borne plant diseases and pathogen antagonists, in Environmentally Sound Agriculture, Frager Press, New York, 51-70.

Sharma, S.K. 1989. Studies on ginger yellows caused by *Fusarium* spp. and its management. M.Sc. Thesis, UHF, Solan. pp. 87.

Vir, D. and R.K. Sharma. 1984. Evaluation of neem oil for control of plant pathogens. *Asian Farm Chem.* **1**(7/8): 23-24.

Neem for the Management of Crop Diseases, 1995
Editor : V. Mariappan
Associated Publishing Co., New Delhi, India
pp. 85-90.

USE OF NEEM IN INTEGRATED MANAGEMENT OF ROT DISEASES OF SMALL CARDAMOM (*ELETTARIA CARDAMOMUM* MATON.)

Dhanapal, K*., Joseph, Thomas and Naidu, R.

Indian Cardamom Research Institute, Myladumpara, Kailasanadu-685 553
P.O. Idukki District, Kerala, India

INTRODUCTION

Rot diseases such as clump rot and capsule rot (Azhukal) of small cardamom (*Elettaria cardamomum* Maton.) are the most serious diseases causing considerable crop loss. These diseases are prevalent both in nurseries as well as in plantations. Rhizome rot and capsule rot (Azhukal) are caused by *Rhizoctonia solani* Koh. and *Phytophthora meadii* Mc RAc respectively. So far the diseases have been managed following plant sanitation coupled with fungicidal applications. However, fungicidal control has got its own limitations such as involvement of high costs, residual problems, environmental hazards etc., In recent years scientists are on the look out of non-chemical means of plant disease control using biocontrol systems or antagonistic plant extracts, which are comparatively less expensive and non-hazardous. Some of the higher plants have recently been shown to be potential sources of fungitoxic substances. Avdhesh Narain and Satapathy (1977) have reported that extracts from *Vinca rosea* inhibited mycelial growth, sporulation and spore germination of the wilt pathogen, *Fusarium oxysporum.* Crude extracts from garlic (*Allium sativum*) has been shown to inhibit the mycelial growth of a number of phytopathogenic fungi. (Shekhawat and Prasada, 1971, Misra and Dixit, 1976). So far no work has been done on the use of antagonistic plant extracts to control cardamom diseases. The present investigation was carried out to evaluate the antifungal properties of different parts of neem, garlic and commercially available neem based formulations against rot pathogens of cardamom under *in vitro* and field conditions.

MATERIALS AND METHODS

***In vitro* studies**: Aqueous extracts prepared from leaves, unripened fruits, and seeds of neem (*Azadirachta indica*), garlic bulbs (*Allium sativum*) and commercially available neem based formulations such as neem oil, Kemisal, Azabin-NSB and Neem Gold were used in the studies. Fresh samples were collected, washed in

* ICRI Regional Research Station, Thadiyankudisai-624 212

tap water and finally in three changes of sterile distilled water. Extracts were prepared in sterile distilled water by grinding the samples in a sterilized mortar and pestle at the rate of 0.5 g/ml on fresh weight basis. The extracts were strained through Whatman No. 1 filter paper. Effect of the plant extracts on the radial growth of *R. solani* and *P. meadii* was studied following poisoned food technique described by Grover and Moore (1962). These extracts were thoroughly mixed with sterilized molten potato dextrose agar to make different concentrations. Fifteen ml. of the medium mixed with the extracts were poured in sterilized petriplates. Petriplates poured with pure PDA alone served as the control. Plates containing the amended and control media were inoculated at the centre with 5 mm. discs of seven day old *R. solani* and *P. meadii* cultures. Inoculated plates were maintained at room temperature (28±3°C). Each treatment was replicated four times. The colony diameter of *R. solani* and *P. meadii* in test and control plates were measured after 120 hours intervals and percentage of growth inhibition was calculated.

Pot culture and Field studies: The *in vivo* effects of plant extracts on *R. solani* infection on six months old seedlings was studied in pot culture under green house conditions. Cardamom seedlings were raised in the nursery and transplanted to earthernware pots (8″ diameter), each pot containing five seedlings. Inoculum of *R. solani* was prepared by grinding 5 days old PDA cultures in sterile water (one plate of culture/100 ml water). In the case of *P. meadii* infection the spray inoculation method (Peethambaran *et al.,* 1988) was followed in six months old seedlings using 7 day old cultures on carrot agar. Zoospores were prepared by the method of Tsao and Garbar (1970). The final spray fluid was prepared by filtering the suspension and adjusting the concentration to get 10^5 zoospores/ml. Seedlings were inoculated by spraying of zoospore (25 ml/seedling) suspension with a glass automiser. Five days later, the test seedlings of *R. solani* infection were drenched and sprayed for *P. meadii* infection with the minimum inhibiting concentrations of respective extracts at the rate of 100 ml/seedlings. The control plants were applied with equal quantity of water. Observations were recorded 15 days after the application of extracts and percent disease incidence on leaves and seedling mortality were calculated.

In Field Conditions: Field evaluation of the antagonistic extracts and neem formulations was carried out in a plot severely affected with clump rot caused by *R. solani.* The trial was carried out using selected concentration (Minimum inhibiting concentration) of respective plant extracts with five replications. Plants were sanitered by removing the disease affected parts before application. The respective plant extracts were drenched at the plant base @ 2 to 3 litres/clump. Observations on disease incidence were taken by counting healthy and infected tillers. The data were analysed statistically.

RESULTS AND DISCUSSIONS

Results on *in vitro* inhibition of *R. solani* and *P. meadii* with plant extracts are presented in Table-1. All the concentrations of the extracts tested inhibited the

Table 1. ***In vitro*** **effect of neem, neem formulations and garlic on mycelial growth of** ***R. solani*** **and** ***P. meadii***

Plant Extracts	Conc. (%)	% Growth inhibition after 120 hours	
		R. solani	*P. meadii*
Neem leaf	5	50.34	45.00
	10	81.37	60.60
	20	86.01	94.23
	30	96.58	100.00
Neem seed	5	50.15	38.42
	10	76.21	100.00
	20	84.99	100.00
	30	100.00	100.00
Neem unripend fruit	5	49.22	25.25
	10	66.00	100.00
	20	71.01	100.00
	30	81.09	100.00
Kemisal	0.1	40.61	65.63
	0.2	82.59	100.00
	0.3	92.15	100.00
Azabin-NSB	0.5	57.81	45.61
	1.0	61.15	87.40
	2.0	96.82	100.00
Neem oil	1.0	50.10	34.60
	2.0	66.53	69.37
	3.0	76.13	100.00
	4.0	100.00	100.00
Garlic bulb	1.0	38.60	44.11
	2.0	51.56	57.45
	3.0	76.46	75.29
	4.0	100.00	90.19

Mean values of four replications

pathogen in varying degrees. Maximum inhibition of mycelial growth was recorded at higher concentrations of the extracts. Growth inhibition started 48 hrs. after inoculation of the pathogen and reached maximum after 120 hrs. No sclerotial formation was observed in any of the treated cultures except in control plates. The concentrations causing 80-100% inhibition were selected as the minimum inhibiting concentrations (MIC) and these were used for testing in the field. Complete inhibition of growth of *R. solani* was observed with extracts from neem seed (30%), neem oil (4%) and garlic bulb (4%). Similar inhibitory properties of plant extracts from neem, garlic and *Bougainvillea spectabiles* on the mycelial growth and sclerotial germination of *Thanatophorus cucumeris*, the perfect state of *R. solani*, has earlier been reported (Lakshmanan and Mohan, 1989). Dubey and Dwivedi (1991) also showed that extract of *A. sativum* was inhibitory to mycelial growth and sclerotial germination of *Macrophomina phaseolina.* In the case of *P. meadii* the mycelial growth was completely inhibited in different concentration of all extracts tested under *in vitro* condition. Root extract of garlic inhibited the germination of chlamidospores and encysted zoospores of *Phytophthora palmivora* in soil and reduce the damping-off of the pappaya seedlings (*Powell and Ko,* 1986). The inhibition of *R. solani* and *P. meadii* observed in the present study using extracts from garlic bulb may be attributed to the presence of inhibitory compounds such as Allicin as reported by Skinner (1955). The active principle i.e., Azadirachtin present in neem products may be responsible for the inhibition of the pathogen.

R. solani inoculated seedlings in the pots showed initiation of disease symptoms 10-15 days after the inoculation of the pathogen. Symptoms were in the form of rotting of leaf sheaths and leaves. Such leaves dried-off at a later stage. In severe cases, the collar portions showed decay symptoms and the entire seedling collapsed. In the case of *P. meadii* sprayed seedlings, water soaked lesions were formed in the leaf and finally rotting occured by the association of the lesions. Results on the extent of foliar infection (Table-2) showed difference between control and treatments. However, significant difference has been observed in mortality rates of seedlings of treated and control pots. Among the neem products tested, it was found that neem seed extract (20%) was most effective in reducing the seedling mortality rates followed by neem oil (3%) and neem unripened fruit (20%). Neem leaf extract, Azabin-NSB and Kemisal also reduced the mortality rates as compared to untreated control. Garlic extract tested (4%) also minimized the disease incidence to a great extent. Leaf infection caused by *P. meadii* was significnatly reduced in neem oil 3% followed by all the extracts tested in this study.

Table 2. Effect of neem, neem formulations and garlic on *R. solani* and *P. meadii* infection in cardamom seedlings

Sl. No.	Treatments	Conc. (%)	*R. solani* infection (%)		Conc. (%)	*P. meadii* infection (%)
			Infection on leaves	Mortality		
1.	Neem Leaf	20	12.37 (20.67)	13.33 (21.09)	20	31.50 (27.33)
2.	Neem Seed	20	15.54 (23.00)	4.00 (9.32)	10	29.94 (24.92)
3.	Neem unripened fruit	20	12.53 (20.70)	8.00 (16.08)	10	32.43 (28.76)
4.	Kemisal	0.3	11.32 (19.65)	16.67 (23.16)	0.3	33.71 (30.82)
5.	Azabin-NSB	2.0	17.16 (24.06)	16.00 (23.19)	2	29.45 (24.24)
6.	Neem oil	3.0	8.13 (15.92)	9.33 (14.33)	3	26.92 (20.52)
7.	Garlic bulb	4.0	27.33 (31.47)	8.67 (16.75)	4	29.03 (23.56)
8.	Control	–	28.67 (32.15)	41.67 (40.00)	–	50.37 (59.20)
			S.E. = 2.17	S.E. = 4.38		S.E. = 0.97
			CD 5% = 6.59	CD 5% = 13.29		CD 5% = 2.95

Figures in paranthesis are transformed values

The effect of plant extracts on clump rot in the field showed encouraging results.

Disease incidence in treated plots were lesser as compared to control plots. Among the extracts tested, Neem Gold (1%) was found to be the best in reducing disease severity. Other extracts were also found to be on par with each other (Table-3). Garlic extract was known to inhibit a number of soil borne fungi and control damping-off disease of several crop plants (*Powell and Ko,* 1986). Application of neem cake and neem leaves in sick soil has been reported to reduce leaf infection of *Collectotrichum capsici* in betelvine (Marimuthu and Rabindran, 1991). Similar inhibitory effects of neem extracts on *R. solani* and their efficacy in controlling damping-off was reported in *Phaseolus aureus* (Lakshmanan and Mohan, 1989). Neem extracts and other neem formulations have been largely used in controlling pest problem of several crop plants (Ahmed and Grainge, 1986 Subrahmanyam, 1990) but are being seldom used in controlling plant diseases. Our study shows encouraging results on the use of neem products in the control of soil borne pathogens. The potentiality of neem and neem based formulations cannot be ruled out in modern disease management strategies though not complete suppression of pathogens is possible.

Table 3. Evaluation of neem and garlic extracts against clump rot caused by *R. solani* in the field

Sl. No.	Treatments	Conc. (%)	Disease incidence (%)	
1.	Neem Leaf	20	10.57	(18.62)
2.	Neem Seed	20	11.52	(16.80)
3.	Neem unripened fruit	20	14.05	(21.76)
4.	Kemisal	0.3	18.89	(25.60)
5.	Azabin-NSB	2.0	10.74	(18.26)
6.	Neem Gold	1.0	9.01	(17.13)
7.	Garlic bulb	4.0	12.33	(20.09)
5.	Control	–	26.46	(30.83)
			L.S.D. (5%) = 7.49	
			L.S.D. (1%) = 10.11	

Figures in paranthesis are Transformed values

ACKNOWLEDGEMENTS

Our sincere thanks are due to Miss. S. Sreelatha and Mr. N. Thulasidas for their help in statistical analysis and typing works. We also gratefully acknowledge M/s. SPIC LTD, Madras for kindly supplying Neem Gold for conducting the trials.

REFERENCES

Ahmed, S. and M. Grainge. 1986. Potential of the neem tree (*A. indica*) for pest control and rural developments. *Econ. Bot.*, **40**: 201.

Avdhesh Narain and J.N. Satapathy. 1977. Antifungal characteristic of *Vinca rosea* extracts. *Indian Phytopath.*, **30**: 36-40.

Dubey. R.C. and R.S. Dwivedi. 1991. Fungitoxic properties of some plant extracts against vegetative growth and sclerotial viability of *Macrophomina phaseolina. Indian Phytopath.*, **44**: 411-413.

Grover, R.K. and C.D. Moore. 1962. Toximetric studies of fungicides against brown rot organisms *Sclerotia fructicola* and *S. laxa. Phytopathology,* **52**: 876-880.

Lakshmanan, P. and S. Mohan. 1989. Antifungal properties of some plant extracts against collar rot of *Phaseolus aureus. Madras Agric. J.,* **76**(5): 266-270.

Marimuthu, T. and R. Rabindran. 1991. Non-chemical method of management of Betelvine anthracnose. *Madras Agric. J.,* **78**: 55-56.

Misra, S.B. and S.N. Dixit. 1976. Comparison of activity of active principle of *Clematis qourina* with commercial fungicides. *Indian Phytopath.,* **29**: 448.

Peethambaran, C.K., R. Suseela Bhai, J. Thomas and R. Naidu. 1988. Standardisation of screening techniques against 'Azhukal' disease of small cardamom (*Elettaria cardamomum* Maton). *J. Plantn. Crops,* **18**(Suppl): 252-256.

Powell, C.R. and W. Ko. 1986. Screening for Antagonistic Plants for Control of *Phytophthora palmivora* in soil. *Ann. Phytopath. Soc. Japan.,* **52**: 817-824.

Shekhawat, P.S. and R. Prasada. 1971. Antifungal properties of some plant extracts, 11 Growth inhibition studies. *Sci. and Cult.,* **37**: 40-41.

Skinner, F. A. 1955. Antibiotics, In: *Modern Method of Plant Analysis,* K. Peach and M.V. Tracey, Vol 111. Springer verlag, Vierter band, pp. 626-725.

Subrahmanyam, B. 1990. Azadirachtin - A naturally occurring insect growth regulator. *Proc. Indian Acad. Sci.* (Ann. Sci.), **99**: 277-288.

Tsao, P.H. and M.J. Garbar. 1970. Methods of soil infestation watering and assessing degree of root infection *in situ* ecological studies with citrus *Phytophthora. Plant Dis. Reptr.,* **44**: 710-715.

Neem for the Management of Crop Diseases, 1995
Editor : V. Mariappan
Associated Publishing Co., New Delhi, India
pp. 91-97.

POTENTIAL OF NEEM AND OTHER PLANT EXTRACTS FOR MANAGEMENT OF CROP DISEASES

Moses, G.J.

Department of Plant Pathology, A.P. Agricultural University, S.V. Agricultural College
Tirupathi-517 502, Andhra Pradesh

INTRODUCTION

That the use of chemicals in plant protection leads to hazards of plant, animal and human health was felt and attempts are made in recent times to replace these with much safer materials. Their value is very much applicable in treatment of post-harvest diseases which can be easily managed by dipping the fruits/bulbs in such material. As a sequel to this, much work was done in our laboratory on management of crop diseases - fungal, bacterial and viral - both *in vitro* and *in vivo* with plant extracts. A review of this work is presented in this article.

MATERIAL AND METHODS

a. Post-harvest diseases: Standard leaf extracts were prepared by grinding one gram of fresh leaf in one ml water and filtered through double layer of muslin cloth. For *in vitro* studies, the standard extract was diluted with PDA (1:10) parts while in *in vivo* studies, it was diluted with water (1:10) parts.

Ether extracts: About 20 kg fresh green leaf was shade dried for 10-12 days (care being taken to prevent fungus development). The dried leaves were ground in a grinder and the material was passed through a 30 mesh sieve to get powder and stored in plastic containers. Petroleum ether (BP 40-60°C) extracts were prepared from the above powder in a Soxhlet apparatus. The samples were dried in a dessicator and hot water bath till a constant weight was obtained. This dried material was considered as 100 per cent crude extract. Stock solutions were prepared from the crude extracts (stored in a deep freezer) with acetone and a sticker as and when required. Further dilutions were made with acetone.

In vitro screening was carried out by Poisoned Food Technique with the test organisms while *in vivo* was by preinoculation dip treatment of the fruits/under wound and unwounded inoculations.

Symptoms were recorded eight days after treatment. Banana fruits were scored in a O to IV grade scale depending upon the percentage of surface area attacked.

Treatment of onion bulbs with plant extracts.

Apparently healthy onion bulbs were surface sterilised, dipped in plant extracts for five minutes and shade dried. They were placed in polythene bags and incubated for 24 hrs at room temp (28 to 32°C). They were dipped in fungal spore suspension for five minutes after wounding with pin cushion and incubated in closed polythene bags at room temp. Percentage rot was estimated by noting the loss in weight of the bulbs.

b. Bacterial diseases: Standard leaf extracts were prepared by grinding one gram fresh leaf material in one ml water and filtered through a Buchner funnel having a double layer of Whatman No. 1 filter paper and sterilised.

Centrifuged extracts were prepared by centrifuging the standard extracts at 30,000 rpm for 30 min. The supernatent was filtered through a double layer of Whatman No. 1 filter paper.

Filter paper disc method was used for testing the efficacy of plant extracs and expressed as area of inhibition zone.

c. Virus diseases: The efficacy of plant extracts in inhibiting virus diseases was studied by pre-inoculation application and the symptoms and incubation were recorded. The standard extracts were prepared by grinding one gram leaf material in five ml water and squeezed through double layered muslin cloth. One part of the leaf extract was mixed with one part of virus sap. The mixture was thoroughly shaken and inoculated on the leaves of the test plants after dusting carborundum powder.

Suitable controls were maintained in all cases.

RESULTS AND DISCUSSION

The results of screening of different plant extracts on pathogenic fungi in post-harvest diseases *in vitro* and *in vivo* are presented in Tables 1 to 3.

Table 1. Percentage inhibition of pathogenic fungi on banana fruits *in vitro*.

S.	Host extract (0.8 p.c.)	*Colletotrichum musae*	*Botryodiplodia theobromae*	*Gloeosporium musarum*	*Fusarium moniliforme*	*Aspergillus niger*	*Rhizopus stolonifer*
1.	Neem	98.68	98.27	96.61	100.00	54.61	98.88
2.	*Bougainvellea spectabilis*	96.05	99.08	87.84	98.82	54.61	97.00
3.	*Parthenium hysterophorus*	98.81	95.40	86.59	97.35	75.64	33.66
4.	*Croton sparsiflorus*	96.71	84.24	77.56	86.76	86.84	14.88
5.	Control	0.00	0.00	0.00	0.00	0.00	0.00

(Seethalakshmi, 1991)

All the plant extracts had considerably (from 86 to 100 p.c.) inhibited the pathogenic fungi on banana studied *in vitro.*

Table 2. Inhibition of *Aspergillus niger* on guava *in vitro.*

S. No.	Plant extract (0.1 p.c.)	Inhibition
1.	Neem	Nil
2.	*Bougainvillea spectabilis*	Partial
3.	*Parthenium hysterophorus*	Partial
4.	*Croton sparsiflorus*	Nil
5.	Control	Nil

(Bharati, 1990)

Extracts of *Bougainvillea* and *Parthenium* partially inhibited *A. niger* on guava while that of Neem and Croton did not do so probably because of the very low concentration 60.1 p.c.) of the extracts tested.

Table 3. Effect of plant extracts on pathogenic fungi on onion bulbs *in vivo.*

S. No.	Plant extract	Percentage decrease over control		
		Aspergillus flavus	*A. niger*	*Rhizopus* sp.
1.	*Aloe vera*	40.10	44.29	39.33
2.	*Calotropis gigantia*	17.70	16.53	42.01
3.	*Cassia auciculata*	27.02	31.60	31.37
4.	*Delonix regia*	0.00	37.60	36.11
5.	*Mangifera indica*	5.26	22.85	22.07
6.	*Moringa oleifera*	4.27	8.37	6.47
7.	Neem	2.71	28.03	27.23
8.	*Thevetia peruviana*	0.00	27.22	24.65
9.	*Prosopis spicifera*	0.00	12.04	10.95
10.	*Leucaena lucocephala*	17.32	22.90	23.20

(Sudhakara Rao, 1987)

Of the ten plant extracts tested *in vivo,* those of *Aloe vera, Calotropis gigantia, Cassia auriculata* and *Delonix regia* had considerable effect while those of Neem, *Mangifera indica, Thevetia peruviana* and *Lruvsrns leucocephsls* had moderate effect and the rest had very low effect in reducing different diseases of onion.

The results of *in vitro* efficacy of plant extracts on the phytopathogenic bacterium, *Xanthomonas campestris* p.v. *citri* are presented in Table 4.

of the 16 plant extracts tested *in vitro,* against *Xanthomonas campestris citri,* the incitant of acidlime canker, crude and centrifuged extracts of neem cake, *Allium sativum, Polyalthia longifolia* and *Lantana camera* had superior efficacy (60 to 137 mm^2 inhibition area) while those of *Croton sparsiflorus Bougainvellea spectabilis, Ficus bengalensis, Datura stramonium, Parthenium hysterophorus,* neem leaf,

Cassia auriculata, Vinca rosea and *Eucalyptus* sp. had moderate effect (25 to 56 mm^2 inhibition area) and the rest had the least inhibitory effect. The ether extracts were found more inhibitory than the crude or centrifuged extracts (80 to 305 mm^2 inhibitory area).

Table 4. Effect of plant extracts on *Xanthomonas campestris citri in vitro.*

S. No.	Plant extract	Mean area of inhibition (mm^2)		
		Crude extract	Centrifuged extract	Ether extract
1.	*Cassia auriculata*	25	31	53
2.	*Ficus bengalensis*	46	42	104
3.	Neem leaf	28	35	131
4.	Neem cake	137	137	—*
5.	*Bougainvillea spectabilis*	54	56	80
6.	*Calotropis gigantia*	13	16	—
7.	*Parthenium hysterophorus*	31	31	84
8.	*Tridax procumbens*	10	12	—
9.	*Tephrosia purpurea*	89	89	—
10.	*Croton sparsiflorus*	56	58	134
11.	*Lantana camera*	62	60	—
12.	*Allium sativum*	134	135	—
13.	*Vinca rosea*	25	28	—
14.	*Datura stramonium*	31	28	—
15.	*Eucalyptus* sp.	25	20	—
16.	*Polyalthia longifolia*	137	131	305
17.	Control	0	0	0

* Not tested (Chandramohan and Moses, 1993)

The results of the studies on the effect of plant extracts on virus diseases as pre-inoculation treatment are presented in Tables 5 and 6.

Table 5. Effectof plant extracts on inhibition of ragi mosaic disease.

S. No.	Host extract	Percentage inhibition	Incubation period (days)
1.	*Achras zapota*	75.00	10.8
2.	Neem	0.00	9.1
3.	*Basella rubra*	91.66	11.8
4.	*Bougainvellea spectabilis*	83.33	11.4
5.	*Catheranthus roseus*	16.67	9.1
6.	*Datura metal*	25.00	9.7
7.	*Delonix regea*	0.00	9.2
8.	*Mirabilis jalapa*	100.00	0.0
9.	*Peltoforum ferrugenium*	75.00	10.8
10.	*Polyalthia longifolia*	0.00	9.1
11.	*Prosopis chilensis*	83.34	11.5
12.	*Tamarindus indica*	58.34	10.5
13.	Control	0.00	9.0

(Swarupa Rani, 1990)

Of the 12 plant extracts tested against ragi mosaic, extracts of *Mirabilis jalapa, Achras zapota, Basella rubra, Bougainvellea spectabilis, Peltoforum ferrugenium* and *Prosopis chilensis* were superior in reducing (83.34 to 100 p.c.) the disease.

Table 6. Effect of plant extracts on inhibition of brinjal mosaic disease

S. No.	Host extract	Percentage inhibition	Incubation period (days)
1.	*Achras zapota*	41.70	15.1
2.	Neem	58.40	15.8
3.	*Bougainvillea spectabilis*	75.00	18.5
4.	*Catheranthus roseus*	16.70	15.4
5.	*Eucalyptus citrodera*	58.40	15.0
6.	*Mirabilis jalapa*	100.00	0.0
7.	*Polyalthia longifolia*	0.00	15.2
8.	*Prosopis chilensis*	83.40	19.8
9.	*Tamarindus indica*	25.0	15.2
10.	Control	0.00	15.2

(Bharati, 1992)

Of the nine plant extracts used against brinjal mosaic, the extacts of *Mirabilis jalapa, Bougainvillea spectabilis,* and *Prosopis chilensis* had superior efficacy while Neem and Eucalyptus extracts had moderate efficacy.

Several workers showed the efficacy of plant extracts in inhibiting the growth of pathogenic fungi *in vitro* as well as in reducing the diseases *in vivo.* Misra and Dixit (1976) observed extracts of *Allium sativum* arrested the growth of *Absidia spinosa, Alternaria tenuis* and *Cephalosporium graminium.* Bhowmick and Varadhan (1981) showed that extracts of *Cinnamomum camphora* and *Catheranthus rosous* inhibited growth, sporulation and spore germination of *Curvularia lunata.* Someswara Rao (1985) found extracts of *Calotropis gigantia* gave 100 p.c. and 95.43 p.c. inhibition of spore germination of *Helminthosporium turcicum* and *Alternaria solani* respectively.

Sheodhan Singh *et al.* (1972) observed aqueous plant extracts of *Datura stramonium, Cannabis sativus, Eucalyptus* sp. *Thuja sinensis* and tobacco were effective at 100 p.c. concentration against hill bunt of wheat when inoculated seed was dipped in plant extracts.

In the present study also, some plant extracts inhibited the pathogenic fungi *in vitro* and onion bulbs and banana fruits were protected from their pathogens when dipped in plant extracts.

Grainge *et al.* (1985) reported that extracts of *Allium sativum* were inhibitory to the phytopathogenic bacterium, *Xanthomonas campestris oryzae.* Hatagalung (1988) observed garlic suppressed the incidence of bacterial wilt of tomato.

In the present study, extracts of neem cake and *Allium sativum* and others were very superior in inhibiting *X. campestris citri.* The ether extracts were found more effective.

Raychaudhury and Prasad (1965) showed inhibitory effect of plant extracts and microbial growth products on infection of raddish mosaic virus. Raychaudhury and Chadda (1965) reported that extracts of deodar fruit inhibited chilli mosaic virus. Gupta and Raychaudhury (1972) noted leaf extracts of *Callistoma lanceolata* and *Syzygium cumini* inhibited PVY. Dhaliwal and Dhaliwal (1971) reported inhibition of TMV on *Phaseolus vulgaris* by extracts of *Allium cepa* and *A. sativum.* Chaudhury and Saha (1988) showed inhibition of urd been leaf crinkle by different plant extracts.

In the present study also, plant extracts, particularly of *Mirabilis jalapa* and others were very effective in reducing the virus diseases tested.

SUMMARY

There is ample evidence that plant extracts are highly useful *in vitro* and *in vivo* in inhibition/reduction of plant pathogens/diseases which can be further exploited by their application particularly on post-harvest diseases in crop disease management.

ACKNOWLEDGEMENTS

The author's grateful thanks are due to P. Sudhakar Rao, V. Bharati, V. Seethalakshmi, B. Chandra Mohan, M. Bharati and G.N. Swarupa Rani whose these formed the source of this review.

REFERENCES

Bharati, M. 1992. Studies on mosaic disease of brinjal (*Solanum melongena* L.) caused by cucumber mosaic virus. M.Sc. (Ag.) thesis submitted to the A.P. Agrl. University.

Bharati, V. 1990. Studies on pre-harvest and post-harvest diseases of guava (*Psidium guajava* L). M.Sc. (Ag.) thesis submitted to the A.P. Agrl. University.

Bhowmick, B.N. and V. Vardhan. 1981. Antifungal activity of some leaf extracts of medicinal plants on *Curvularia lunata. Indian Phytopath.,* **34**: 385-386.

Chandra Mohan, B. and G.J. Moses. 1993. Effect of extracts of neem and other plants on *Xanthomonas campestris citri,* the incitant of citrus canker. World Neem Conference, Bangalore, India, pp. 41. (Abst.).

Chaudhury, A.K. and N.K. Saha. 1988. Inhibition of urd bean leaf crinkle virus by different plant extracts. *Indian Phytopath.,* **38**: 366-368.

Dhaliwal A.S. and G.K. Dhaliwal. 1971. Inhibition of TMV multiplication by extracts from *Allium cepa* and *A. sativum. Advancing Frontiers of Plant Sciences,* **28**: 305-310.

Grainge, M., L. Berger, and S. Ahmad. 1985. Effect of extracts of *Artabotrys uncinatus* and *Allium sativum* on *Xanthomonas campestris oryzae. Curr. Sci.,* **54**: 90.

Gupta, V.K. and S.P. Raychaudhury. 1972. Inhibition of PVY by the leaf extracts of *Callistemon lanceolatus* and *Syzygium cumini. Indian Phytopath.,* **25**: 108-112.

Hatagalung, L. 1988. Garlic bulb as a material for suppressing the incidence of bacterial wilt (*Pseudomonas solanacearum*) on tomato. *Bull. Penelition Horticulture,* **16**: 184-193.

Misra, S.B. and S.N. Dixit. 1976. Fungicidal spectrum of the leaf extract of *Allium sativum. Indian Phytopathy.,* **29**: 448-449.

Raychaudhury, S.P. and K.C. Chadda. 1965. Deodar fruit extracts as inhibitor of chilli mosaic virus. *Indian Phytopath.,* **18**: 97-98.

Raychaudhury, S.P. and H.C. Prasad. 1965. Effect of plant extracts and microbial growth products on the infectivity of raddish mosaic virus. *Indian J. Microbiol.,* **5**: 13-16.

Seethalakshmi, V. 1991. Studies on post-harvest diseases of banana. M.Sc. (Ag.) thesis submitted to the A.P. Agrl. University.

Sheodhan Singh, L.B. Goel, S.K. Sharma and S.K. Nayar. 1972. Fungitoxicants and plant extracts in the control of hill bunt of wheat. *Indian Phytopath.,* **25**: 297-298.

Someswara Rao, J. 1985. Screening of certain natural products for their fungicidal properties. M.Sc. (Ag.) thesis submitted to the A.P. Agrl. University.

Sudhakara Rao, P. 1987. Storage diseases of onion (*Allium cepa* L.) and garlic (*Allium sativum* L.). M.Sc. (Ag.) thesis submitted to the A.P. Agrl. University.

Swarupa Rani, G.N. 1990. Studies on mosaic disease of ragi (*Eleusine coracana* Gaertn.). M.Sc. (Ag.) thesis submitted to the A.P. Agrl. University.

Neem for the Management of Crop Diseases, 1995
Editor : V. Mariappan
Associated Publishing Co., New Delhi, India
pp. 99-106.

"NEEM FOR MANAGEMENT OF CROP DISEASES"

Bansal, R.K. and Tak, S.K.

Rajasthan Agricultural University, SKN College of Agriculture, Jobner-303 329
District Jaipur (Rajasthan)

INTRODUCTION

Neem has recently attracted the Agricultural Scientists for its usefulness as a pesticide and its ability to provide a pollution free protection to plants.

Extensive use of plant protection chemicals and sometimes indiscriminate use has led to serious social and environmental repercussions. The poisoning of livestock, fish, wildlife and other beneficial organisms has been due to increased use of these chemicals. There has also been a disturbing increase in human poisoning, particularly in developing countries where safe handling and application of insecticides is not always feasible due to several socioeconomic factors.

In the last two decades, interest has increased in the potential and possible use of new bioactive products and natural pesticides, which are less likely to cause ecological damage. Plants virtually are the richest source of bioactive organic chemicals on earth. Although, only about 10,000 secondary plant metabolites have been chemically identified, the total number of plant chemicals may exceed 400000 (Swan, 1977). The chemical or pesticide approach had its beginnings in the use of botanical materials. This reviews the potential of need, *Azadirachta indica* A. Juss (Family Meliaceae) for management of crop diseases.

Chemical composition: Jacobson (1980 and 1986), Kraus *et al.* (1980 and 1986) and Mofan (1980) has reviewed the chemistry of neem compounds. The "bitters" have been identified as limonoids, a group of stereochemically homogenous tetranortriterpenoids. The most important active participat is azadirachtin, although more than 25 different compounds have been isolated including betasitosterol, fatty acids and flavonoids.

Meliantriol and Salannin have been obtained discrete crystalline form and structurally defined, but there has been much inconsistency about the chemical structure of azadirachtin since till recently, it could be obtained only as a white amorphous powder that melts at 154-158°C.

Azadirachtin belongs to the C-seco-limonoid group of alkaloids and occurs only in the need tree or Indian lilac (Ley, Denholm and Wood, 1993).

Products and potentiality: The complexity of neem compounds precludes their synthetic production in near future. Consequently, most neem derivatives used or evaluated for pathogen control include oil cake, aqueous or organic solvent extracts of seed kernel, partially prufied fractions, azadirachtin-rich formulations etc.

The thick, viscuos, crude neem oil has a strong garlic odour due to the presence of a large number of sulfur moieties. India produces approximately 83,000 tonnes of neem oil and about 3,32,000 tonnes of neem cake annually (Ketkar, 1976; Saxena, 1986 and 1987). The major outlet for neem oil in India is the soap industry where it is used in the manufacturer of low grade soaps and detergents. It is also used as a lubricant and as a paint thinner.

India is marketing two neem based formulations-Repelin and Wellgro-for spraying in tobacco-growing areas of Andhra Pradesh to check damage by insect-pests and tobacco mosaic virus in tobacco nurseries (Subramaniam, T.S., I.T.C. Ltd., Rajahmundry, India). Use of Repelin is also becoming popular in cotton growing regions in Andhra Pradesh. An Indian company (Khandal, V.S., West Cost Herbochem, Bombay, India) has came with an azadirachtin-rich granular formulation Neemark, which is recommended for use (at a rate of 18.75 kg/ha) on cotton, paddy, tobacco, groundnut, sugarcane, chilli, eggplant, other common vegetables, horticultural crops, legumes and foodgrains.

Important biological properties of chemical, extracts and preparation from parts of neem have also been highlighted by Koul, Isman and Ketkar (1990). *In vitro* and *in vivo* studies on efficacy of neem against viral, fungal bacterial and nematode pathogens have been conducted.

Neem in the control of fungal pathogens: Number of neem derivatives such as oils, extracts and cakes have been found to be effective against fungal pathogens *in vitro* and *in vivo*.Singh and Pandey (1967) carried out experiments to determine the effects of decomposition of oil cakes, used as organic manures in India, on the population of *Pythium apahanidermatum* in soil. Margosa cake *(Azadirachta Indica)* was highly stimulatory to the fungus in all concentrations of 0.1 to 5 per cent at 20°C - 30°C, but caused inhibition at 36°C. Singh (1968) conducted a trial on Silt-loam-soil in a field in which the potato crop was affected with black scurf last year. The organic matter content of the soil was 1.95 per cent, nitrogen 0.05 per cent and pH 6.2. They used uniformly infected seed tubers of almost identical size of variety on 1645, highly susceptible to the disease. Five different cakes were mixed in soil at the rate of 2500 kg/ha. The amendment of soil with oil cakes best control was provided by mustard cake with disease incidence of 12 per cent followed by peanut (16 per cent), Margosa (25 per cent) and castor (25 per cent) cake over check (31 per cent). Mahua cake enhanced disease incidence and reduced yield of tubers. Fungitoxicity test of neem leaves extract was carried out by Bambode and Shukla (1973). The autoclaved and unautoclaved extract did not produce any inhibition (mm) against *Curvalaria lunata, Alternaria tenuis, Helminthosporium sativum, H. speciferum* and *Rhizoctonia botaticola* but showed stimulatory effect on growth of *Fusarium moniliforme.* Neem oil cake is reported to inhibit the conidial germination of *Sclerospora* (*Peronosclerospora*) *sacchari.* In the glasshouse *Azadirachta* oil and cake gave promising results. A combination of chloroneb seed treatment, roguing and 1 foliar spray of azadirachta oil gave the most effect control of *S. sacchari* in maize (Lal and Saxena, 1980). Singh *et al.* (1980) suggested possible role of *A. indica* extracts and oil in the control of soil borne pathogens of gram (*Cicer arietinum*). Growth of *Fusarium oxysporum* of sp. *ciceri, Rhizoctonia solani, Sclecotium rolfsii* and *Scleoctinia sclerotiorum* was inhibited in liquid medium by extracts of leaf, trunk, bark, fruit pulp and oil of neem (*A. indica*). Higher concentrations of oil inhibited seed germination. Oil treated seeds sown in soil infested with the pathogens, singly or mixed, produced disease free seedlings whereas, all the seedlings from untreated seeds exhibited disease symptoms. The fruit pulp suppressed the formation of sclerotia by *R. solani,* concentrations of extracts and oils from neem (100-5000 ppm) was found to effect adversely the germination and germ tube growth of *Erysiphe polygoni* (DC) conidia (Singh and Singh, 1981). Inhibitory effect of extracts of *Azadirachta indica* was reported on the mycelial growth and spore germination of *Curvularia lunata* and *Alternaria alternata* (Bhowmick and Vardhan, 1981). Different concentrations of neem oil have found to inhibit the growth in number of fungi (Vir and Sharma, 1985). At concentration of 2.5 and 5 per cent radial growth of *Fusarium moniliforme* (Gibberella fujikuroi) was inhibited by 57.8-66.7 per cent; that of *Aspergillus niger* by 77.8%; that of *Drechslera rostratum* by 83.3 -88.9%; and that of *Macrophomina phaseolina* by 61.1 - 75.6%. At 10.0 per cent the oil gave cent percent inhibition of all the test pathogens. At 1.0 per cent concentration growth of *Alternaria alternata* was inhibited by 61.1 per cent whereas, at 10 per cent it was inhibited by 100 per cent. Dwivedi and Dubey (1986) observed more deleterious effects of volatile fraction (VF) of hydrodistillates of *Azadirachta indica* on germination of *Macrophomina phaseolina* sclerotia than non-volatile fractions (NVF). A notable decrease in sclerotial germination due to VF and NVF of neem oil and neem oil cake extracts was observed after 24d incubation. Muthuswamy, *et al.* (1988) observed neem seed, and cake extracts and neem oil to be highly effective in inhibiting spore germination of *Puccinia arachidis,* causing rust of groundnut. Leaf oil of *A. indica* showed fungitoxic activity against *Sclerotium rolfsii* (Singh *et al.,* 1989). Eswaramurthy and co-workers (1989) found inhibition in the mycelial growth of *Sarocladium oryzae* and *Fusarium oxysporium* f. sp. *cepae* due to extracts of *A. indica,* Ghewande (1989) recorded effective control of *Puccinia arachidis* and *Phaeoisariopsis personata* (*Mycosphaerella* berkeleyii) when groundnuts were sown and treated with aqueous leaf extract of *Azadirachta indica,* resulting in 15.42 - 39.84 per cent increase in the yield. Germination of purified *sclerotia* of *Slerotium* (*Corticium*) *rolfsii, Rhizoctonia* (*Ceratobasidium*) *Oryzae sativae* and *Sclerotium hydrophyllum* was strongly inhibited when treated with 5% plant oil emulsion of *A. indica* (Banerjee *et al.,* 1989). Water extracts of *A. indica* significantly inhibited mycelial growth and sclerotial germination of *Thanatephorus cucumeris* under green house conditions (Lakshman and Mohan, 1989). Bansal and

Sobti (1990) suggested that *A. indica* is a cheaper and readily available approach in the control of crop diseases. They found least per cent disease incidence of (2.66%) of *Aspergillus niger* on groundnut when seeds were treated with neem leaf extract (2%). It was significantly superior over check (24%) and thiram (0.3%) and dithane M-45 (0.2%). Neem extract was most effective and significantly increased the seed germination (76%), followed by thiram and dithane M-45 over check (53.33%). Thiram, asafoetida, neem extract and dithane M-45 showed significantly less seedlings mortality over check (59.48%). These were also found to increase significantly the radicle and plumule length of groundnut seed over check.

According to Pan (1990) in slide germination and growth inhibition tests with *Drechslera oryzae* (*Cochliobolus miyabeanus*), the addition of *Azadirachta indica* oil to captan increased the fungicidal activity synergistically but had the reverse effect with carbendazim. Upadhyay and Rai (1990) found 70 per cent inhibition in the mycelial growth of *Curvularia fubuculata,* (causal agent of citrus die-back) when treated with leaf extract of *A. indica.* Mishra and Tewari (1990) observed that ethanolic extract of leaves of *A. indica* effectively inhibited *Pyricularia oryzae conidia* and mycelial growth *Rhizoctonia solani* and *Fusarium moniliforme* mycelial growth was also found to be completely inhibited. Rice plants tested were found to possess principles against one or the other tested fungal pathogens. Neem oil is reported (Ganapathy and Narayanaswamy, 1990) to reduce the incidence of late leaf spot (*Phaeoisariopsis personata*), rust (*Puccinia arachidis*) and ring mosaic (tomato spotted wilt virus) diseases of groundnut. In addition, pod yield was also found to be increased by 62.3 per cent.

The antifungal activity of leaf extract of neem, *Azadirachta indica* on wheat, at different dilution (1:2, 1:4, and 1:8), showed marked reduction in seed mycoflora as well as enhancement in seed germination (Khan and Shah, 1993). Sarvamangala *et al.* (1993) found significant reduction in uredospore germination of *Cerotelium fici* in all the plant extracts in study. Maximum inhibition of spore germination was observed in leaf extract of *Azadirachta indica* (91.2%). They also observed 18.9 per cent control of the leaf spot disease of mulberry, caused by *Cercospora moricola,* when sprayed with extract from fresh leaves of *A. indica* under field conditions.

Solid amendment with neem oil cake is also known to have pronounced effect on rhizosphere mycoflora. According to Goswami and Bhattacharya (1989), in unsterilized soil, more frequency of fungi were encountered than sterilized amended soil. Fungi population was found to be increased during decomposition of neem and groundnut oil cakes attributed to the release of acids and other chemicals which change the soil environment in favour of dormant fungi. Reduction in microbial population and activity in soil was noticed when neem coated phorate granules wer applied (Babtiwale and Dethe, 1989).

Neem as remedy for control of plant viral diseases: Extracts of *Azadirachta* *indica* was found to be most potent in reducing virus infectivity (Tripathi and Tripathi, 1982).

Crude oil from margosa (*A indica*, as a 0.5 per cent water emulsion inhibited the transmission of CMV (Cucumber) to an extent of 40.0, 33.3, 23.3 and 30.0 per cent when the aphids were allowed to come into contact with the oil before *acquition* during acquisition was recorded when plants were sprayed with oil emulsion after inoculation. Oil did not effect the biological activity of the virus as evidenced by a mechanical transmission test using a mixture of oil and virus. The mechanism of inhibition in relation to aphid transmission of CMV seems to be different from that reported for mineral oils (Srivastava *et al.,* 1986). Zaidi *et al.* (1988) observed inhibition of spinach mosaic virus by *A. indica* leaves. Leaf and bark extracts of *A. indica* inhibited infection of *Chenopodium amaranticolor* by cowpea mosaic virus. The inhibitor was heat stable, non-dialyasable through cellophane, remained active after 25 days storage, was not destroyed by desication for 50 days, showed appreciable loss of activity on dilution, was destroyed by activated charcoal and induced localised resistance. Antiviral activity was lowered in strong acidic and basic solutions and was reversed by simulataneous application of Actinomycin D. The extract was very effective when applied to the leaves prior to inoculations (Singh *et al.,* 1989).

Leaf extracts of *Azadirachta indica* reporded low infection (40%) of rice yellow dwarf as compared to extracts of other test plants. The incubation period of the virus increased to 40.3 days as against 28.5 days in control. Neem oil treatment also recorded lower per cent infection (13.3%) as compared to control (82.2%) and also significantly increased the incubation period (Narasimha Rao and Narayanaswamy, 1990).

Reduction of virus transmission: The use of oil formulations with antifeedants from plants for controlling insect-transmitted viral diseases was emphasized by Simons (1981). Saxena *et al.* (1980) observed that the incidence of *Nilaparvata lugens*-transmitted ragged stunt virus disease was significantly less in rice fields periodically sprayed with 12 per cent neem oil than in unsprayed field. Neem oil reduced the *N. lugens* survival and suppressed the transmission of grassy and ragged stunt virus diseaes (Saxena and Khan, 1985). Neem oil alone or mixed custard apple oil also reduced the survival of *N. virescens and its transmission of rice tungro viruses on treated seedlings (Mariappan and Saxena, 1983 and 1984).*

Neem cake (5%) and Kernel extract (2%) is also known to reduce the incidence of yellow mosaic virus transmitted by whitefly Bemisia tabaci in black gram (Mariappen *et al.,* 1987).

The incidence of virus diseases was generally lower in rice fields applied with a neem cake urea mixture than in the fields applied with urea only (Saxena *et al.,* 1984). Enzyme linked immunosorbent assays showed that rice seedlings grown in soil treated with neem cake at 150 kg/ha had a significantly lower incidence of infection of rice tungro bacilliform and sphericol viruses transmitted by *N. virescens* than in control seedlings (Saxena *et al.*, 1987). Protection with neem cake at 250 kg/ha was comparable with that provided by application of carbofuran 3G at 0.75 kg (AI)/ha.

Future scope and prospects: The present review reveals the significant effect of neem derivates against plant pathogens. Although 250 limnoids, including azadirachtin, from Meliaceous plants have been identified (Taylor, 1984), more may be discovered. The use of neem bitters and other bioactive fractions, rather than neem oil as such will ease the competition with the soap industry, which at present is a major buyer of neem oil. The semipurified neem seed bitters and other neem-rich fractions can easily be standarized for biological properties and can satisfy even stringent quality requirements. Being water soluble, they can also be applied as systemic compounds, which renders them more photostable and nonphytotoxic, unlike the neem oil. Their efficacy can further be enhanced with additives, synergists, and antioxidants, and improved methods of application.

More neem trees will have to be grown to meet the increasing demand for industrial, fungicidal and insecticidal uses. Selection of superior ecotypes will be desirable to increase the productivity and quality of neem produce. Once established, neem trees will be good sources of raw materials for industrial, insecticidal and fungicidal use even for hundreds of years.

The safety of neem derivatives to predators and parasities and other non-target organisms, including man is now well established. In this context, the use of neem derivatives offers a harmonious and non-violent approach to get plant disease management. Also, the neem compound's ability to inhibit nitrification, should make their use in crop husbandry doubly attractive.

To benefit fully, the cost of neem based pesticides must remain within the user's reach. To achieve this, farmers should be encouraged to grow neem on their home steads and be educated about its values in plant disease management.

ACKNOWLEDGEMENT

We thank Dr. A.K. Sanghi, Professor and Dean, SKN College of Agriculture, Jobner for useful comments and help in the preparation of this review.

REFERENCES

Babtiwale, H.P. and Dethe, M.D. (1989) Effect of neem coated phorate granules on microbial respiration in soil. Neem Newsl. **6**(3): 25-27.

Bambode, R.S. and Shukla, V.N. (1973). Antifungal properties of certain plant extracts against some fungi. The Punjab Krishi Vidyapeeth Research Journal **2**(1): 1-8.

Banerjee, S., Bhattacharya, I. and Mukherjee, N. (1989). Sensitivity of three sclerotial rice pathogens to plant oils. Intn. Rice. Res. Newsl. **14**: 23.

Bansal, R.K. and Sobti, A.K. (1990). An economic remedy for the control of two species of *Aspergillus* on groundnut. Indian Phytopath. **3**(43): 451-452.

Bhowmick, B.N. and Vardhan, V. (1981). Antimycotic activity of leaf extract of some medicinal plants on *Curvularia lunata.* Indian Phytopath. **34**: 385-386.

Dwivedi, R.S. and Dubey, R.C. (1986). Effects of volatile and non-volatile fractions of two medicinal plants on germination of *M. phaseolina* sclerotia. Trans. Brit. Mycol. Soc. **87**: 326-328.

Eswaramurthy, S., Muthuswamy, M., Muthuswamy, S. and Mariappan, V. (1989). Inhibitory effects of neem, Acacia, Prosopis and Ipomea extracts on *Sarocladium oryzae* and *Fusarium oxysporum* f. sp. *cepae.* Neem Newsl. **6**(1): 4-5.

Ganapathy, T. and Narayanasamy, P. (1990). Effect of plant products on the incidence of major disease of groundnut. Intn. Arachis Newsl. **7**: 20-21.

Ghewande, M.P. (1989). Management of foliar disease of groundnut (*Arachis hypogea*) using plant extracts. Indian J. Agri. Sci., **59**: 133-134.

Goswami, B.K. and Bhattacharya, D. (1989). Efficacy of neem and groundnut oil cakes as influenced by micro-organisms changes in rhizosphere mycoflora during decomposition of oil cakes in sterilized and unsterilized soil. Indian J. Nema. **19**(1): 72-74.

Jacobson, M. (1980). Proc. Ist Inst. Neem Conf., Rottach Egern. pp. 33-42.

Jacobson, M. (1986). Proc. 3rd Inst. Neem Conf., Nairobi, 33-44.

Jacobson, M. (1986). *In Natural Resistance of Plants to Pests: Roles of Allelochemicals* (Eds., Green, M.B. and Hedin, P.A., 220-232.

Khan, M.J. and Shah, N. (1993). Antifungal activity of leaf extract of neem seed on mycoflora of wheat. **3**(2): 209-210.

Koul, O., Isman, M.B. and Ketkar, C.M. (1990). Properties and uses of neem (*Azadirachta indica*). Canadian J. Bot. **68**(1): 1-11.

*Kraus, W., Cramer, R., Bokel, M. and Switzki, G. (1980). Proc. Ist. Intn. Neem Conf., Rottach Egern., 53-62.

*Kraus, W., Baumann. S., Bokal, M., Keller, U., Klenk, A. Klingele, M., Pohns, H. and Schwinger, M. (1986). Proc. 3rd Intn. Neem Conf., Nairobi, 111-125.

Lakshmanan, P. and Mohan, S. (1989). Antifungal properties of some plant extracts against collar rot of *Phaseolus aureus.* Madras Agri. J., 266-270.

Lal, S., Nath, K. and Saxena, S.C. (1980). Use of pesticides and natural products in control of *Sclerospora sacchari* in maize. Tropical Pest Management, **26**: 286-292.

Ley, S.V., Denholm, A.A. and Wood (1993). The chemistry of Azadirachtin. Natural Product Reports **10**(2): 109-157.

Mariappan, V. and Saxena, R.C. (1983). Effect of custard-apple oil and neem oil on survival of *Nephotettix virescens.* J. Econ. Entomol. **76**: 573-576.

Mariappan, V. and Saxena, R.C. (1984). Effect of mixtures of custard apple oil and neem oil on survival of *Nephotettix virescens* and on Rice Tungro virus transmission. J. Econ. Entomol. **77**: 519-521.

Mariappan, V., Gopalan, M., Narasimhan, V. and Suresh, S. (1987). Neem Newsl. **4**(1): 9-10.

Mishra, M. and Tewari, S.N. (1990). Ethanolic extract toxicity of three botanicals against five fungal pathogens of rice. National Acad. Sciencel. **13**(11): 409-412.

Morgan, E.D. (1980). Proc. 2nd Intn. Neem Conf., Rottach Egern. 43-52.

Muthuswamy, M., Eswaramurthy, S., Muthuswamy, S. and Mariappan, V. (1988). Evaluation of neem products against rust disease of groundnut. Neem Newsl. **5**(4): 48.

Narasimha, Rao, G. and Narayanaswamy, P. (1990). Effect of plant extracts and oils on rice yellow dwarf infection. Madras Agri. J. **77**: 197-201.

Pan, S. (1990). Additive effects of plant oils and fungicides. J. Mycopathol. Res. **28**: 57-61.

Sarvamangala, H.S., Govindaiah, and Datta, R.K. (1993). Evaluation of plant extracts for the control of fungal diseases of mulbery. Indian Phytopath. **46**(4): 398-401.

Saxena, R.C. (1986). Proc. 3rd Intn. Neem Conf., Nairobi, 81-93.

Saxena, R.C. (1987). In Pesticide Science and Biotechnology. (Eds. Greenhalgh, R. and Roberts, T.R.). Blackwell Scientific, London, 139-144.

Saxena, R.C., Liquido, N.J. and Justo, Jr. H.D. (1980). Proc. 1st Intn. Neem Conf., Rottach-Egern. 171-188.

Saxena, R.C., Justo, Jr. H.D. and Epino, P.B. (1984). Evaluation and utilization of neem cake against the rice Brown plant hopper. *Nilaparvata lugens.* J. Econ. Entomol. **77**: 502-507.

Saxena, R.C. and Khan, Z.R. (1985). Effect of neem oil on survival of *Nilaparvata lugens* and on grassy stunt and ragged stunt virus transmission. J. Econ. Entomol. **78**: 647-651.

Saxena, R.C., Khan, Z.R. and Bajet, N.B. (1987). Reduction of tungro virus transmission by *Nephotettix virescens* in neem cake treated rice seedlings. J. Econ. Entomol. **80**: 1079-1082.

Simons, J.N. (1981). In vectors of Disease Agents, Interaction with plants, Animal and Man. (Mc Kelvy, Jr. J.K., Elridge, B.F. and Maramarasch, K.) Praeger, N.Y. 169-178.

Singh, A.K., Manju Singh and Singh, A.K. (1988). Antiviral activity and physical properties of extracts of *Azadirachta indica*. Indian J. Virology, **4**: 76-81.

Singh, H.B. and Singh, U.P. (1981). Effect of volatiles of some plant extracts and their oils on conidia of *Erysiphe polygoni* DC. Aust. Pl. Pathol. **10**: 66-67.

Singh, R.K., Shikla, R.P. and Dwivedi, R.S. (1989). Studies on fungitoxicity of oil against *Sclerotium rolfsii* Sacc. and soil mycoflora. National Acad. Sciencel. **12**(6): 183-185.

Singh, R.S. and Pandey, K.R. (1967). Population dynamics of *Pythium aphanidermatum* in oil cake amended soils. Canadian Journal of Microbiology **13**: 601-610.

Singh, R.S. (1968). Incidence of Black scurf on potato in oil cake amended soil. Indian Phytopath. **21**(1): 120-121.

Singh, U.P., Singh, H.B. and Singh, R.B. (1980). The fungicidal effect of neem (*Azadirachta indica*) extracts on some soil borne pathogens of gram (*Cicer arietinum*). Mycologia, **72**: 1077-1093.

Srivastava, K.M., Rama, M.S., Dwadash Shreni, V.C. and Singh, B.P. (1986). Machanism of inhibition of CMV by crude oil from margosa (*A. indica*) during inoculation with single apterous *Aphis gossypi*. Indian Phytopath. **39**: 20-25.

*Swain, T. (1977). Ann. Rev. Plant Physiol. **28**: 479-501.

Taylor, D.A.H. (1984). In progress in the chemistry of organic natural products. (Eds. Herz, W., Grisebach, H. and Kirby, G.W.). Springer-Verlag, Vienna, **45**: 1-102.

Tripathi, R.K.R. and Tripathi, R.N. (1982). Reduction in bean common mosaic virus (BCMV) infectivity vis-vis crude leaf extract of some higher plants. Experentia, **38**(3): 349.

Upadhyay, S. and Rai, M.K. (1990). *In vitro* sensitivity of *Curvularia-tuberculata* Jain: A causal organism of dieback disease of citrus to different plant extracts. Indian Medicine. **1**: 13-15.

Vir, D. and Sharma, R.K. (1985). Efficacy of fungicides XXXII. Evaluation of neem oil for the control of plant pathogens. Asian Farm Chemicals. **1**: 23-24.

Vir. D. and Sharma, R.K. (1985). Efficacy of fungicides XXIX. Studies on the fungicidal properties of neem oil. Indian J. Pl. Pathol. : 241-242.

Zaidi, Z.B., Gupta, V.P. and Samoda, Naqvi, Q.A. (1988). Inhibition of spinach mosaic virus by extracts of some medicinal plants. Current Science. **57**: 151-152.

Neem for the Management of Crop Diseases, 1995
Editor : V. Mariappan
Associated Publishing Co., New Delhi, India
pp. 107-114.

EFFECT OF APPLICATION OF PLANT PRODUCTS ON THE HOST PLANT RESISTANCE

Manickam, K. and Mariappan, V.

Department of Plant Pathology, Tamil Nadu Agricultural University
Coimbatore-641 003, Tamil Nadu

Protection of crop plants from the ravages of pests and diseases by the use of synthetic pesticides has been the usual practice for many years. In the present days constraints in pest control, especially those imposing chemical pesticides due to their ecological hazards, development of pesticide resistance and their expensive nature etc., botanical pesticides are gaining stature and recognition as a possible method of practical control of diseases of crop plants. Plant derivatives possessing pesticidal properties are offering greater scope than the other group of materials.

Plant derivatives were found to contain an array of chemicals and these chemicals were reported to induce various types of influences on the pathogen as well as on the host plant. Kamalakkannan (1994) observed the extracts of plant species were effective in reducing the rice blast disease caused by *Pyricularia oryzae.* The plant extracts contained several bio-chemical constituents (Table 1). Among the plant extracts analysed, the reducing, non-reducing and total sugar contents were relatively more in the leaf extracts of *Boerhaavia diffusa* followed by *Lantana camara, Azardirchta indica* and *Abutilon indicum* than that of others. Maximum protein content was observed in *A. indica* followed by *Prosophis juliflora* and *Zizyphus jujuba. P. juliflora* was found to contain maximum amount of glycoprotein followed by *Z. jujuba, A. indicum* and *A. indica.* The total phenol content was maximum in *Z. jujuba* followed by *A. indicum* and *P. juliflora. Z. jujuba* was possessing maximum tannin content followed by *P. juliflora* and *A. indica* (Table 1). Among these plant extracts *P. juliflora* has been highly effective in reducing the growth of the pathogen and it contained maximum amounts of glycoprotein, protein, total phenol and tannin.

Jayalakshmi (1994) analysed the components of the plant species that were found to be effective in reducing the chilli mosaic virus disease (Table 2). The total sugar and non-reducing sugar contents were maximum in *Lantana camara* followed by neem seed kernal extract (NSKE). The reducing sugar content was more in *L. camara* followed by *Bougainvillea spectabilis.*

Table 1. Bio-chemical constituente of plant extracts which wear against rice blast pathogen

Sl. No.	Leaf extracts of plant species*	Percent reduction on the mycelial growth of the fungus (1)	Total sugars mg/g (2)	Reducing sugar mg/g (3)	Non-reducing sugar mg/g (4)	Protein content mg/g (5)	Glycoprotein content mg/g (6)	Total phenol content mg/g (7)	Tannin content mg/g (Dry weight) (8)
1.	*Abutilon indicum*	78.1 $(62.1)^{c}$	10.2^{d}	6.9^{c}	3.3^{e}	15.3^{h}	1.5^{b}	2.9^{b}	23.5^{d}
2.	*Aclaypha indica*	64.1 $(53.2)^{f}$	7.4^{f}	4.6^{g}	2.8^{f}	31.6^{a}	0.7^{f}	1.9^{f}	12.6^{f}
3.	*Azadirachta indica*	66.8 $(58.8)^{f}$	12.4^{c}	6.6^{d}	5.8^{b}	25.5^{cd}	1.1^{d}	0.8^{h}	26.5^{c}
4.	*Boerhoavia diffusa*	67.8 $(55.4)^{h}$	15.7^{a}	10.4^{a}	5.3^{c}	21.5^{a}	0.8^{ef}	0.7^{f}	14.5^{gh}
5.	*Clerodendron inerme*	72.9 $(58.7)^{f}$	6.8^{f}	4.3^{h}	2.5^{g}	17.9^{efgh}	0.8^{e}	2.2^{e}	18.9^{f}
6.	*Cynodon dactylon*	77.8^{d} (61.9)	8.4^{g}	6.8^{c}	1.6^{h}	20.4^{ef}	1.0^{d}	1.3^{g}	18.1^{f}
7.	*Cyperus rotundus*	74.1^{e} (59.4)	7.9^{h}	5.2^{f}	2.6^{f}	18.7^{efg}	0.9^{e}	2.7^{c}	17.0^{g}
8.	*Lantana camara*	69.3^{g} (56.3)	15.6^{b}	8.4^{b}	7.2^{a}	26.4^{cd}	1.2^{c}	2.5^{d}	22.2^{e}
9.	*Prosophis juliflora*	93.3^{a} (75.0)	9.4^{f}	6.2^{e}	3.2^{e}	29.8^{ab}	1.9^{a}	2.9^{b}	30.0^{b}
10.	*Zizyphus jujuba*	82.0^{b} (64.9)	9.6^{e}	5.4^{f}	4.2^{d}	28.5^{abc}	1.8^{a}	4.5^{a}	33.5^{a}

Concentraction 10%
All values in the first column are n percentages
Values in parentheses are arcsine trnsformed values
In column means followed by same letter (s) are not significantly different by DMRT (P = 0.05)

Kamalakkannan, 1994.

The amino acid and protein content were maximum in NSKE followed by *B. spectabilis* and *L. camara. C. inerme* contained maximum amount of glycoprotein (Table 2).

The phenol content was more in *B. spectabilis* followed by *L. camara* and *A. indicum, where as the tannin content was more in Casuarina equisetifolia* followed by *A. indica.*

However the NSKE containing maximum amount of amino acid and protein content has been found to be slightly more effective than *L. camara* or *B. spectabilis* in reducing the chilli mosaic virus diseasse. The plant extracts when applied to the susceptible infected host plants or on the healthy plant, were found to change the bio-chemical activities, in the treated plants.

Verma (1988) reported that application of the extracts of non-host plants such as *B. diffusa, Clerodendron aculeatum, C. fragrans* and *Pseudoranthemum bicolor* on susceptible plants like tobacco, tomato, urdbean, soyabean and sunnhemp induced systemic resistance in these plnats towards subsequent virus infection. Induction of resistance was found host mediated. Development of resistance was prevented when metabolic inhibitors like actinomycin-D and cycloheximide were given along with the treatment. Active inhibitory fraction suggested that perhaps macro molecules were involved. These were low molecular weight and chemical tests were consistent with macro molecules such as polypeptides, glycoproteins or proteins. These biologically active fractions were shown to be highly water soluble low molecular proteins. Under stimulating action of these proteins, non circulatory virus inhibitory agents were synthesized *de-novo* in the genetically susceptible cultivars of plants. Virus inhibitory agents were also low molecular weight highly soluble proteins and their formation involves either gene de-repression or activation of a latent enzyme system. These induced protein provided adequate defence to plants to infection by viruses. The virus inhibitory agents are thus plant products elicitor host metabolites. They possess virus neutralizing property and play a major role in the inhibition of virus *in-vivo.* Regular sprays at suitable intervals with these plant products provided protection to crop plants against virus infection and also improved the growth and yield.

Application of plant extract preparation have been reported to interfere with enzyme activities of susceptible host plants. Verma and Vivek Prasad (1987) reported AVP from *Clerodendron aculeatum* induced resistance to Sunnhemp rosette virus in *Cyamopsis tetragonoloba* and this resistance was due to increased activity of catalase, peroxidase (PO) and polyphenol oxidase (PPO). Similar changes in *Nicotiana tabacum* cv. samsun NN inoculated with TNV were reported by Prasad (1986). Increased phenylalanine ammonialyase (PAL) activity leads to the *de-novo* synthesis of phenolics (Glazzener, 1982). PAL is an enzyme of phenyl propanoid pathway and have a definite role in the defence mechanism of plants (Friend and Threshfall, 1976). Peroxidase (PO) is an important enzyme in the synthesis of lignins. It also catalyses the oxidation of phenols to quinones which are highly

Table 2. Bio-chemical constituents of plant extracts used against chilli mosaic virus

Sl. No.	Leaf extracts of plant species*	Percent disease reduction over control (1)	/ug/g of fresh tissues							
			Total sugars (2)	Reducing sugar (3)	Non-reducing sugar (4)	Amino acids (5)	Protein (6)	Glyco-protein (7)	Total phenol (8)	Tannin (9)
1.	*Azadirachta indica* (Neem seed kernal extract, NSKE)	66.6 (55.1)^{a}	12.5^{bcdef}	6.6^{bc}	5.9^{daf}	5.5^{d}	28.4^{f}	1.3^{a}	2.6^{ab}	26.3^{e}
2.	*Lantana camara*	56.6 (48.9)^{a}	15.5^{f}	8.4^{f}	7.1^{f}	2.7^{ab}	26.6^{cdef}	1.4^{ab}	2.7^{abc}	22.7^{ch}
3.	*Abutilon indicum*	50.0 (45.0)^{a}	10.2^{abcdef}	6.8^{bed}	3.4^{abcde}	2.7^{ab}	26.3^{cd}	1.7^{bc}	2.7^{abc}	23.5^{ed}
4.	*Bougainvillea spectabilis*	53.3 (47.0)^{a}	9.9^{abcdef}	7.2^{def}	2.7^{abcde}	3.4^{c}	25.0^{c}	1.4^{ab}	3.8^{d}	16.4^{a}
5.	*Casuarina equisetifolia*	60.0 (50.9)^{a}	9.2^{abcde}	6.5^{bc}	2.7^{abcd}	2.7^{ab}	15.5^{a}	1.7^{bc}	2.6^{ab}	31.5^{f}
6.	*Clerodendron inerme*	50.0 (45.0)^{a}	6.9^{a}	4.3^{a}	2.6^{ab}	1.1^{a}	17.6^{ab}	2.3^{d}	2.4^{a}	19.5^{ab}
	C.D.	12.91								

* Concentration 10% — Jayalakshmi, 1994.

In column one values in parentheses are arcsmine transformed values.

In column means followed by same letters are not significantly different ($P = 0.05$) by DMRT fungi toxic. Peroxidase itself is toxic to some pathogens. It has antibacterial activity (Vidhyasekaran, 1992). PPO oxidizes phenols to highly fungi toxic quinones. The enzyme itself is inhibitory to some plant viruses. It inactivates RNA of the virus (Mink and Saksena, 1981). Kamalakkannan (1994) applied the plant extracts on rice plants (IR50) which is susceptible to blast disease. The bio-chemical changes occurring in the blast infected rice plants that were applied with the plant products either before inoculation of the pathogen (*Pyricularia oryzae*) or after the inoculation of the pathogen was compared with the healthy plants. The results obtained have revealed many changes occuring after the application with the plant products.

Effect of pre-inoculation treatment on rice blast infected plants: The PO, PPO and PAL activities were reduced in blast infected rice plants when compared to healthy rice plants. However the PO, PPO and PAL activities were enhanced in plants that had been treated with plant extracts of *P. juliflora* and *Z. jujuba* considerably than those observed in rice plants infected by blast pathogen as well as that of healthy rice plants (Table 3).

The total phenol content was found reduced when the rice plants infected by blast pathogen. But the phenol content increased significantly when the rice plants treated with plant extracts, compared to healthy rice plants. *P. juliflora* enhanced the phenol content to an extent o 3.00 mg/gm of leaf tissues followed by *Z. jujuba* compared to 2.6 mg/gm of healthy leaf tissues. *A. indica* (neem), *A. indicum* and *C. dactylon* plant extracts increased the phenol content to a level of 2.9 to 2.8 mg/gm of fresh tissues in rice plants (Table 3).

Table 3. Application of plant products as pre-inoculation spray against *Pyricularia oryzae*

Sl. No.	Leaf extracts of plant species	PO activity	PPO activity	PAL activity	Total phenol mg/g of fresh tissues
1.	Control (Healthy untreated)	64.0^{f}	6.0^{f}	8.1^{f}	2.6^{de}
2.	Control (Infected untreated)	42.0^{g}	3.7^{g}	6.4^{g}	1.8^{f}
3.	*Abutilon indicum*	96.0^{d}	8.9^{bc}	12.1^{bc}	2.8^{cd}
4.	*Azadirachta indica*	99.3^{c}	8.7^{cd}	12.0^{cd}	2.8^{cd}
5.	*Cynodan dactylon*	95.0^{e}	8.0^{de}	11.2^{e}	2.8^{cd}
6.	*Prosophis juliflora*	103.3^{a}	9.1^{ab}	12.6^{ab}	3.0^{a}
7.	*Zizyphus jujuba*	100.7^{b}	9.8^{a}	13.1^{a}	2.9^{ab}

Kamalakkannan, 1994

0.01 = OD = 1 Unit for PO, PPO and PAL activity.
In column means followed by same letter(s) are not significantly different (P = 0.05) by DMRT.

Effect of post inoculation treatment on rice blast infected plants: Kamalakkannan (1994) reported that the PO, PPO and PAL activities were less in blast infected rice plants when compared to healthy plants. But PO and PAL activities were found to increased significantly in these plants when treated with leaf extracts of *P. juliflora, Z. jujuba* and *A. indica* followed by other treatments when compared to uninoculated healthy and that of blast infected rice plants (Table 4). The PPO activity also enhanced slightly when the rice plants were treated with *A. indica* and other plant products when compared to healthy and control rice plants.

The total phenol content was decreased when the rice plants infected by *P. oryzae*. But the total phenol content was increased significantly when the plants were treated with *A. indica* (neem) and *P. juliflora* when compared to healthy as well as blast infected rice plants (Table 4).

Table 4. Application of plant products as post-inoculation spray against *Pyricularia oryzae*

Sl. No.	Leaf extracts of plant species	PO activity	PPO activity	PAL activity	Total phenol mg/g of fresh weight
1.	Control (Healthy untreated)	71.0^{f}	5.1^{d}	7.3^{e}	2.6^{e}
2.	Control (Infected untreated)	62.0^{g}	3.0^{e}	2.0^{f}	2.0^{f}
3.	*Abutilon indicum*	114.3^{d}	6.6^{c}	14.4^{c}	3.2^{a}
4.	*Azadirachta indica*	117.7^{c}	8.6^{a}	13.8^{d}	3.2^{a}
5.	*Cyanodon dactylon*	110.3^{e}	6.4^{c}	13.8^{d}	2.8^{c}
6.	*Prosophis juliflora*	123.0^{a}	8.0^{b}	15.3^{b}	3.1^{b}
7.	*Zizyphus jujuba*	120.0^{b}	8.0^{b}	15.5^{a}	2.7^{c}

Kamalakkannan, 1994

0.01 = OD = 1 Unit for PO, PPO and PAL activity.
In a column means followed by same letter(s) are not significantly different (P = 0.05) by DMRT.

Effect of pre-inoculation of plant extracts on the incidence of tomato spotted wilt virus on mungbean plants: The catalase, PO and PAL activities were remarkably increased when the plants were treated with plant extracts (AVPs), the maximum being in plants treated with *Croton sparsiflorus* (Table 5). Inoculation of plant extracts treated plants with TSWV reduced the activity of these enzymes which however was always higher than in healthy and inoculated plants (Manickam, 1991). The protein content increased marginally over healthy plants when plants treated with plant extracts. Plants treated with plant extracts and inoculated also showed marginal increase in total protein content, but it was significantly less than that in inoculated plants (controls). Marginal increase in protein content in plant extract treated plants may be formation of additional proteins. These proteins involved in the development of resistance.

Chilli mosaic virus infection has been reported to reduce the application of plant extracts (Duleepkumar Samuel 1990 and Jayalakshmi, 1994). Bio-chemical changes occuring in chilli plants due to the application of leaf extracts has been described by Jayalakshmi (1994).

The PAL, PO, PPO and Ascorbic acid oxidase (AO) activities were found increased in chilli mosaic virus infected plants when compared to healthy. But these activities were found further enchanced when the chilli plants were treated with the plant

Table 5. Bio-chemical changes in mungbean plants due to pre-inoculation spray of plant products against mungbean leaf curl disease

Sl. No.	Treatments	OD / min / mg			Total soluble protein (mg/g)
		Catalase	PO	PAL	
1.	Healthy	0.256^{e}	0.029^{c}	0.123^{d}	138.33^{b}
2.	Control (inoculated)	0.272^{d} (6.25)	0.035^{b} (20.68)	0.129^{c} (4.87)	141.41^{a} (2.22)
3.	Coconut AVP treated	0.277^{bc} (8.20)	0.036^{ab} (24.13)	0.132^{b} (7.31)	138.65^{b} (0.231)
4.	Treated + Inoculated	0.276^{c} (7.81)	0.036^{ab} (24.13)	0.132^{b} (7.31)	138.7^{b} (0.267)
5.	Sorghum AVP treated	0.278^{ab} (8.59)	0.037^{a} (27.38)	0.133^{ab} (8.13)	138.71^{b} (0.274)
6.	Treated + Inoculated	0.277^{bc} (8.20)	0.037^{a} (27.38)	0.132^{b} (7.31)	138.7^{b} (0.310)
7.	*Prosophis chilensis* AVP treated	0.278^{ab} (8.59)	0.037^{a} (27.38)	0.133^{ab} (8.13)	138.73^{b} (0.289)
8.	Ttreated + Inoculated	0.277^{bc} (8.20)	0.037^{a} (27.38)	0.133^{ab} (8.13)	138.7^{b} (0.310)
9.	*Croton sparsiflorus* AVP treated	0.279^{a} (8.98)	0.037^{a} (27.38)	0.134^{a} (8.94)	138.73^{b} (0.289)
10.	Treated + Inoculated	0.279^{a} (8.98)	0.037^{a} (27.38)	0.134^{a} (8.94)	138.73^{b} (0.289)
		CD:0.00135	CD:0.00137	CD:0.00186	

Manickam, 1991

Data in parentheses are percentage increase (or) decrease over healthy
Incolumns means followed by same letters are statistically on par (P = 0.05)

extracts (Table 6). The chilli plants treated with NSKE recorded maximum activity than the other treatments.

Table 6. Bio-chemical changes in chilli plants due to foliar spray of plant extracts followed by chilli mosaic virus inoculation

Sl. No.	Leaf extracts of plant species	PAL activity	PO activity	PPO activity	Ascorbic acid oxidase	Protein ug/g
1.	*Abutilon indicum*	0.40^{bcde}	0.10^{cdef}	0.04^{bcde}	0.33^{def}	466.6^{f}
2.	*Azadirachata indica* (NSKE)	0.50^{efg}	2.23^{g}	0.06^{fg}	0.35^{ef}	480.0^{eg}
3.	*Bougainvillea spectabilis*	0.43^{bcdef}	0.07^{abcde}	0.04^{bc}	0.31^{de}	383.3^{cde}
4.	*Casuarina equisetifolia*	0.40^{bcd}	0.05^{abcd}	0.04^{cde}	0.28^{d}	369.3^{cd}
5.	*Clerodendton inerme*	0.38^{bc}	0.04^{abc}	0.03^{ab}	0.18^{b}	275.3^{a}
6.	Chilli plants infected with chilli mosaic virus	0.30^{ab}	0.03^{ab}	0.04^{bcd}	0.18^{bc}	336.7^{bc}
7.	Healthy chilli plant	0.23^{a}	0.23^{a}	0.02^{a}	0.10^{a}	310.3^{ab}

Jayalakshmi, 1994

* Mean of the values between 0-30, 30-60, 60-90 seconds Mean separation in a column by DMRT at 5 per cent level.

The NSKE was found to be highly effective than the other plant species in reducing the chilli mosaic virus disease. Although *B. spectabilis* enhance the enzyme activity it was not as effective as neem in reducing the chillimosaic virus disease. Similarly *A. indicum* possessing high protein also was not as effective as NSKE in reducing the chilli mosaic virus disease. But the NSKE alone was found to enhance both protein content as well as the enzyme activities which are the precursors of phenol metabolism. And probably the combined action of protein and the influences of the enzymatic activities were responsible for the effectiveness of NSKE in reducing chilli mosaic virus. From these observation it is clearly revealed that plant products when applied to the susceptible host plants were able to interfere with the bio-chemical activities of the host plants which were helpful in reducing the disease severity.

REFERENCES

Duleepkumar Samuel, L. 1990. Studies on chilli mosaic virus. M.Sc. (Ag.) Thesis, TNAU, Madurai. pp. 133.

Friend, J. and D. Threshfall. 1976. Bio-chemical aspects of Plant parasitic relationship. Academic press, London, pp. 354.

Glazzener, J.A. 1982. Accumulation of phenolic compounds in the physiology of plant disease and disease resistance *Phytopath. Z.,* **44**: 105-150.

Jayalakshmi, V. 1994. Management of chilli mosaic virus by plant products. M.Sc. (Ag.) Thesis, TNAU, Coimbatore. pp. 142.

Kamalakkannan, A. 1994. Management of rice blast (*Pyricularia oryzae*) disease by using chemicals and plant extracts. M.Sc. (Ag.) Thesis, TNAU, Coimbatore. pp. 201.

Manickam, K. 1991. Management of greengram *Vigna radiata* (L.) WILCZEK leaf curl disease using Antiviral principles and chemicals. M.Sc. (Ag.) Thesis, TNAU, Coimbatore. pp. 132.

Mink, G.I. and K.N. Saksena. 1971. Studies on the mechanism of oxidative inactivation of plant viruses by O-quinones. *Virology,* **45**: 755-763.

Prasad, V. 1986. Alteration in enzyme activity during induced antiviral state by leaf extract. *J. Indian Bot. Soc.,* **65**: 90-94.

Verma, H.N. 1988. Management of virus diseases with plant products by activating defence in susceptible plants. "National seminar on management of crop diseases with plant products/biological agents". Agrl. College & Research Institute, Madurai. pp. 55. (Abst.).

Verma, H.N. and Vivek Prasad. 1987. Systemic induced resistance by plant extracts alters physiology of susceptible test host. *Indian J. Plant Pathol.,* **5**: 69-72.

Vidhyasekaran, P. 1992. Principles of Plant Pathology, CBS Printers and Publishers, Delhi. 106 pp.

Neem for the Management of Crop Diseases, 1995
Editor : V. Mariappan
Associated Publishing Co., New Delhi, India
pp. 115-121.

EFFICACY OF PLANT PRODUCTS FOR THE MANAGEMENT OF BACTERIAL PLANT PATHOGENS

Narasimhan, V., Selvam, R. and Mariappan, V.

Department of Plant Pathology, Tamil Nadu Agricultural University
Coimbatore-641 003, India

INTRODUCTION

Antibiotics and chemicals are generally used for the control of bacterial plant pathogens. Excessive use of antibiotics lead to the development of acquired resistance to bacteria. Use of broad spectrum antibiotics also lead to the killing of non-target bacterial organisms. To avoid such environmental hazards, alternate sources have to be found out. Many plant species have been reported to contain antibacterial activity and this source can be well utilized for management of Bacterial diseases.

EFFICACY OF PLANT EXTRACTS/PRODUCTS

In-vitro evaluation of Plant extracts/products: All parts of the plant viz., root, bulb, rhizome, leaf, flower, seeds, trunk etc., were found to have antibacterial activity against many bacterial plant pathogens (Table 1).

Table 1. Effect of extracts of different plant parts on bacterial plant pathogens

Sl. No.	Name of the plant part	Name of the plant	Pathogens	References
1.	Entire plant	*Acalypha indica*	*Pseudomonas aeruginosa*	Hirenath *et al.* (1993)
2.	Rhizome	*Curcuma zedoaria*	*Xanthomonas malvacearum*	Anup Banerjee *et al.* (1978)
3.	Bulb	*Allium sativum*	*X. campestris* pv. *vesicatoria*	Grainge and Alvarez (1987)
4.	Trunk	*Quercus acuta*	*P. aeruginosa*	Muney Serit *et al.* (1991)
5.	Leaf	*Artabotrys hexapetalus*	*X. campestris* pv. *campestris*	Grainge and Alvarez (1987)
6.	Seeds	*Moringa oleifera*	*X. campestris* pv. *campestris*	Grainge and Alvarez (1987)

Leaf extracts of *Artabotrys hexapetalus* was tested against 12 phytopathogenic bacteria by Grainge and Alvarez (1987) (Table 2).

Table 2. Inhibitory effect of leaf extract of *Artabotrys hexapetalus* on bacterial plant pathogens

Sl. No.	Bacteria	Diameter of inhibition zone (in mm)
1.	*Agrobacterium tumefaciens*	13
2.	*Clavibacter michiganense*	22
3.	*C. flaccumfaciens*	8
4.	*Erwinia carotovora*	14
5.	*E. herbicola*	3
6.	*Pseudomonas syringae* pv. *phaseolicola*	17
7.	*P. solanacearum*	19
8.	*Xanthomonas campestris* pv. *campestris*	16
9.	*X. campestris* pv. *oryzae*	29
10.	*X. campestris* pv. *phaseoli*	13
11.	*X. campestris* pv. *vesicatoria*	15
12.	*X. campestris* pv. *vitians*	11

Among them *Xanthomonas campestris* pv. *oryzae* was highly inhibited by the leaf extract of *A. hexapetalus.*

Janina Mosch *et al.* (1990) studied the efficacy of plant extracts against fire blight pathogen, *Erwinia amylovora* (Table 3).

Table 3. In-vitro activity of plant extracts against *E. amylovora*

Sl. No.	Plant extract	Inhibition zone (mm)	Minimum inhibition concentration (%)
1.	*Rhus typhina*	28.0	1.3
2.	*Juglans nigra*	28.2	5.0
3.	*Berberis vulgaris*	27.2	2.0
4.	*Ruta graveolans*	25.5	15.0
5.	*Populus tremula*	24.4	15.0
6.	*Sedum groenlandicum*	23.5	10.0
7.	*Pimpinella saxifraga*	23.0	20.0
8.	*Ailanthus altissima*	20.9	20.0
9.	*Castanea sativa*	20.5	7.5
10.	*Polygonum capitatum*	20.0	20.0
11.	*Berberis lampergiana*	20.0	7.5
12.	*Salvia officinalis*	19.5	15.0
13.	*Salvia sclarea*	28.0	19.2
14.	*Berberis vulgaris*	19.0	10.0
15.	Streptomycin (17 ppm)	27.2	2.0

The inhibitory potential of plant extracts from *Juglans nigra, Berberis vulgaris* and *Rhus typhina* applied as aqueous solution in 5, 2 and 1.25 per cent was comparable with the antibiotic streptomycin (17 ppm).

Various plant extracts taken from different parts of plant were evaluated against *Xanthomonas campestris* pv. *campestris* (Grainge and Alvarez, 1987) (Table 4).

Table 4. Inhibitory effect of crude plant extracts on *Xanthomonas campestris* pv. *campestris*

Sl. No.	Name of the Plant	Plant Part used	Diameter of inhibition zone (in mm)
1.	*Allamanda cathartica*	Leaves	0
2.	*Artabotrys hexapetalus*	Leaves	12
3.	*Conocarpus erectus*	Leaves	10
4.	*Eucalyptus citriodora*	Leaves	10
5.	*Lagestromia speciosa*	Leaves	0
6.	*Tulbaghia violacea*	Leaves	5
7.	*Allium sativum*	Bulb	14
8.	*Moringa oleifera*	Seeds	10
9.	*Tamarindus indica*	Pod pulp	13

Among them the bulb extract of *Allium sativum* and leaf extract of *Artabotrys hexapetalus* were highly effective in inhibiting the pathogen.

Bulb extract of *Allium sativum* has been evaluated against 12 bacterial plant pathogens (Grainge and Alvarez 1987) (Table 5).

Table 5. Effect of *Allium sativum* bulb extract on bacterial plant pathogens

Sl. No.	Bacteria	Diameter of inhibition zone (in mm)
1.	*Agrobacterium tumefaciens*	2
2.	*Clavibacter michiganense*	17
3.	*C. flaccufaciens*	13
4.	*Erwinia carotovora*	12
5.	*E. herbicola*	5
6.	*Pseudomonas syringae* pv. *phaseolicola*	18
7.	*P. solanacearum*	12
8.	*Xanthomonas campestris* pv. *campestris*	14
9.	*X. campestris* pv. *oryzae*	29
10.	*X. campestris* pv. *phaseoli*	12
11.	*X. campestris* pv. *vesicatoria*	15
12.	*X. campestris* pv. *vitians*	2

Among the different bacteria tested by them, *Xanthomonas campestris* pv. *oryzae* was highly inhibited by the bulb extract of *A. sativum.*

Trunk extract of *Quercus acuta* (Extracted in ethanol) was reported to have antibacterial activity against gram negative bacteria *Pseudomonas aeruginosa* (Muney Serit *et al.* 1991).

The sequentially partitioned extract of this plant, 4, 5-di-o-galloyl (+) protoquercitol was found to be the active constituent involved in the activity.

Tea and coffee wastes were also tested for their effectiveness against three bacterial plant pathogens. Coffee waste showed strong hypersensitive reaction against *Pseudomonas syringae* var. *pisi* race 1 and 2 (Sadhana Álstrom, 1992).

Different solvents have been tested for extraction from the leaf of *Acalypha indica* against *Pseudomonas aeruginosa* and among them petroleum ether solvent had the maximum inhibitory effect (Hiremath *et al.* 1993).

In-vitro evaluation of essential oils: Essential oils from the leaves of various plant species were tested against three bacterial plant pathogens viz., *Pseudomonas mangifera indica, P. solanacearum* and *Xanthomonas campestris* by Siva Sankara rao and Nigam (1978) and among them *Cinnamomum zeylanicum, Cymbopogon flexuosus* and *Eucalyptus citriodora* were shown to be most effective oils against these three pathogens.

Five essential oils were evaluated by Sampurna and Nigam (1979) against cotton bacterial pathogen viz., *Xanthomonas malvacearum* and they found that *Eucalyptus citriodora* oil had the maximum inhibitory activity (Table 6).

Table 6. Inhibition zone of different doses of essential oils against *X. malvacearum*

Sl. No.	Essential oil of	Concentration tested					
		100%	80%	60%	40%	20%	10%
1.	*Skimmia laureola*	31	28	19	12	12	–
2.	*Cymbopogon flexuosus*	29	23	20	18	17	–
3.	*Cinnamomum zeylanicum*	21	19	18	16	5	13
4.	*Geranium*	22	21	18	14	12	8
5.	*Eucalyptus citriodora*	34	21	28	22	20	11

The effectiveness of essential oil from *Acorus calamus* against some gram negative and gram positive bacteria (Alankara Rao and Rajendra Prasad, 1981), dithym olysed oil of *Carum copticum* against six potato pathogenic bacteria, essential oil of *Chrysanthemum indicum* and *Feronia elephantum* against *Xanthomonas vesicatoria* and *Pseudomonas putida* respectively (Pandey *et al.* 1981) have been proved. These volatile oils being non-phytotoxic can be used as fumigants for protecting against rotting of potato. The essential oil from the leaves of *Anacardicum occidentale* was also effective against *pseudomonas mangifera* and *Xanthomonas campestris* (Garg and Kasera, 1984).

The effectiveness of essential oil of *Cyathocline lyrata* against *Xanthomonas citri, Pseudomonas aeruginosa* and *Bacillus subtilis* (Rajendra Shrivastava *et al.* 1980), of *Glossocardia bosvallia* against *X. campestris, B. subtilis* (Anupam Pathak and

Dixit, 1984), of *Buddleia asiatica* against *X. campestris, P. mangiferae, P. aeruginosa* and *B. subtilis* (Garg and Oswal, 1981) have also been reported.

MANAGEMENT OF BACTERIAL DISEASES

Seed treatment: Seed treatment of cabbage seeds by soaking the seeds in the leaf extract of *Artabotrys hexapetalus* for various periods was evaluated against *Xanthomonas campestris* pv. *campestris* (Grainge and Alvarez, 1987) (Table 7).

Table 7. Effect of *Artabotrys hexapetalus* leaf extract on germination and percentage of infected cabbage seeds (Internally borne *X. campestris* pv. *campestris*)

Sl. No.	Seed treatment	Treatment time	Germination (%)	Infection (%)
1.	*Artabotrys hexapetalus*			
	Crude leaf extract	30 min	94	6
	- do -	3 hour	20	0
2.	Sterile crude extract	30 min	99	2
	- do -	3 hour	7	0
3.	Crude extract + Water (1:9)	30 min	99	7
	- do -	3 hour	95	2
4.	Soxhlet extract	30 min	99	66
	- do -	3 hour	92	7
5.	Boiled fresh leaves	3 hr	81	6
6.	Boiled dry leaves	3 hr	84	7

Crude leaf extract treatment soaked for 30 min reduced the percent of infection as well as increased the per cent germination.

FOLIAR APPLICATION

Foliar treatment under field condition have also been tesed and found effective for some bacterial plant pathogens. Citrus canker caused by *Xanthomonas campestris* pv. *citri,* was effectively controlled by neem cake (22.5 kg) soaked in of water (90 l) kept for decomposition for a week and then sprayed on the foliage. This proved effective not only against pathogen but also controlled the leaf miner, *Phylocnstis citrella.*

ANTIBACTERIAL PRINCIPLE

Identification of antibacterial principle lant products: Although several plant pathogens have been screened against bacterial plant pathogens specific components responsible for antibacterial activity have not been elucidated. However, for some of the plant products the antibacterial principle have been identified (Table 8).

Table 8. Antibacterial principle compound in plant extracts

Sl. No.	Plant extract	Principle	References
1.	*Cinnamomum zeylanicum* *Skimmia laureola* *Cymbopogon flexuosus*	Thymol menthol	Sampurna and Nigam (1979)
2.	*Cinnamomum tamala*	α-phellanderene and Eugenol	Anonyum (1950)
3.	*Boswellia serrata*	d-α-thugene, d-α-pinene, d-limonene	Bhargava (1963)
4.	*Acorus calamus*	Etharial oil	Alankara Rao and Rajendra Prasad (1981)
5.	*Quercus acuta*	4-5 di-o-galloyl protoquercitol 3, 5 di-o-galloyl protoquercitol and gallic acid	Muney Serit *et al.* (1991)
6.	*Ocimum sanctum, O. canum* *O. basilicum*	Methyl chamicol, Citral, Eugenol, Methyl cinnamate linalool	Sinha and Gulati (1990)
7.	*Allium sativum*	Sulphur	Mangamma and Sreeramula (1991)
8.	Tea and Coffee wastes	Tannic acid	Toda (1984)

Asthana *et al.* (1988) reported the antimicrobial activity of essential oils. Since the skeletal constituents of essential oils are terpenoids, six selected terpenoids viz., Citral, Citronellol, Citronellal, Eugenol, Farinisol and Nerol were evaluated against six bacterial plant pathogens (Table 9).

Table 9. Antibacterial activity of some terpenoids against potato pathogenic bacteria

Sl. No.	Name of the Bacterial Pathogen	Diameter of zone of inhibition (in mm)		
		Control	Eugenol	Nerol
1.	*Bacillus polymixa*	–	15	20
2.	*Erwinia carotovora* var. *atroseptica*	–	25	45
3.	*E. carotovora* var. *carotovora*	–	45	25
4.	*Pseudomonas putida*	–	30	28
5.	*P. solanacearum*	–	40	7
6.	*Xanthomonas vesicatoria*	–	40	26

Among the six terpenoids tested, eugenol and nerol were effective against all the bacteria. Terpenoids are non-toxic to plants and highly volatile and will leave no residual problem.

FUTURE THRUST

Evaluation of plant extracts and essential oils from plants have shown promising results against bacterial plant pathogens. Hence further screening and identification of many plant species against important pathogens have to be continued and effectivenes have to be brought into commercial form. Field level testing with botanical formulations have to be investigated further. Combination of botanicals with bactericides for effective control also needs further research. Thus the use of botanicals for control of bacterial plant diseases have vast scope and bright future for effective management without polluting the environment.

REFERENCES

Anonym, 1950. The wealth of India. C.S.I.R. Publication. pp. 178.

Alankara Rao, G.S.J.G. and Y. Rajendra Prasad. 1981. Antimicrobial property of *Acrorus calamus* Linn. Invitro studies. *Indian perfumer.* **XXV**(3 & 4): 46.

Anupbanerjee, V., K. Kaul and S.S. Nigam. 1978. Antimicrobial efficacy of the essential oil of *Curcuma zedoaria* (Rósc) Roxi. *Indian Perfumeer.* **XXII**(3): 214-217.

Anupam Pathak and V.K. Dixit. 1984. Antifungal and antibacterial studies on essential oil of *Glossocardia bosaval. Indian Perfumer* **28**(2): 91-94.

Asthana, A.S. Ahmad, A. Dikshit and S.N. Dixit. 1983. Toxicity of some terperoids against potato pathogenic bacteria. *Indian Perfumer,* **27**(3 & 4).

Bhargava, P. 1963. *Perfumory essential records.* **54**(11): 740.

Garg, S.C. and V.B. Oswal. 1981. Invitro antibacterial activity of the essential oil of Buddleia asiatica lour. *Indian Perfumer,* **XXV**(3 & 4): 7-9.

Garg, S.C. and H.L. Kasera, 1984. Antibacterial activity of the essential oil of *Anacardium occidentale* Linn. *Indian Perfumer.* **28**(2).

Grainge, M.D. and A.M. Alvarez. 1987. Antibacterial and antifungal activity of *Artabotrys hexapetalus* leaf extracts. *Int. J. Tropical Plant diseases,* **5**: 173-179.

Hiremath, S.P., Shrishallappa Damaml, H.K.S. Swamy and D.C. Biraddar. 1993. Antimionbial activity of various extracts of *Acalypha indica, Indian J. of Microbiology.* **33**(1).

Janina Mosch, F. Klingant and W. Zeller. 1990. The effect of plant extracts against Fireblight. *Acta Horticulture.* **273**: 355-359.

Mangamma, P. and A. Sreenramulu. 1991. Garlic extract inhibitory to growth of *Xanthomonas campestris* pv. *vesicatoria. Indian Phytopathology,* **44**: 372-374.

Muney Serit, Tsutomuokubo, Rong-Huisu, Nobuyuki Hagiwara, Muno Kim, Tetsuo Iwagawa and Takehiko Yamamoto. 1991. Antibacterial compounds from Oak. *Quercus Acuta.* Thunb. Agri. Chem. **55**(1): 19-23.

Pandey, O.K.A. Asthana, N.N. Tripathi and S.N. Dixit. 1981. Volatile plant products vis-a-vis potato pathogenic bacteria. *Indian Perfumer. Pastno.* **XXV**(3 & 4): 10-14.

Rajendran, Shrivastava, Ashok Omray and K.C. Varma. 1980. Microbial growth inhibition by essential oil of cyathocline lyrata cass. *Indian Perfumer.* **XXIV**(4): 119-221.

Sadhana Alstrom. 1992. Antibacterial activity of tea and coffee. wastes against some plant pathogenic pseudomonas strains. *J. Phytopathology* **136**: 329-334.

Sampurna, J. and S.S. Nigam. 1979. Anti bacterial activity of some Indian essential oils. *Indian Perfumer.* **XXIII**(3 & 4): 205-207.

Sinha, G.K. and B.C. Gulati. 1990. Antibacterial and antifungal study of some essential oils and some of their constituents. *Indian Perfumer.* **4**(2): 26-129.

Sivasankara Rao, T. and S.S. Nigam. 1978. Antibacterial study of some Indian essential oils. *Indian Perfumer.* **XXII**(2): 118-119.

Toda, M.S. Okuba, R. Hiyoshi, and T. Shimamura. 1989. The bactericidal activity of tea and coffee. *Letters in applied Microbiology,* **8**: 123-125.

B

Management of Diseases Caused by Viruses

Neem for the Management of Crop Diseases, 1995
Editor : V. Mariappan
Associated Publishing Co., New Delhi, India
pp. 125-136.

NEEM BASED PESTICIDES, POTENTIAL AGENTS FOR THE MANAGEMENT OF VIRAL DISEASES OF VEGETABLE CROPS

Singh, S.J. and Krishna Reddy, M.

Division of Plant Pathology, Indian Institute of Horticultural Research
Hessaraghatta Lake P.O. Bangalore-560 089, Karnataka

India is the second largest producer of vegetables next only to China. The estimated area under vegetables during the 7th plan was 5.5 millions hectares with the annual production of 53.8 million tonnes which contributes roughly 12% of the world production of vegetables. During the Eighth Plan period, production targets have been projected at 75 million tonnes from an area of 6 million hectares with annual growth rate of 7.2%.

Vegetables play an important role in the human diet and provide lot of nutrition. A number of vegetable crops are cultivated throughout the country in different agro-climatic zones. Vegetable crops are prone to more than 100 viral and mycoplasma-like diseases. Some of these diseases are of economic importance causing serious losses in yield and quality of the produce. Few economically important viral and mycoplasma diseases occurring in India and their insect vectors have been listed in Table-1.

The present chapter deals with the management of viral diseases of vegetable crops by use of neem-based pesticides.

Neem (*Azadirachta indica*) a native species of the Indo-Burma region, is distributed in India, Pakistan, Bangladesh, Sri Lanka, Burma, Thailand, Malaysia, Indonesia and planted extensively in the middle East and tropical region of Africa. Neem tree has been held in great esteem in India from the time immemorial and references on the diverse uses of different parts of this plant are found in ancient, religious literatures. According to reports from various sources there are about 13.9 million neem trees distributed all over India. Neem tree has a great potential to save environment in the tropics particularly in India.

Scope of neem in pest management: Neem (*Azadirachta indica* A. Juss) known as Indian Lilac is indigenous to India and has become the focus of attention all over world due to its plethora of biolocal activities. These include excellent insect control activities and various medicinal properties. Intensive search during the past decade for a safer insecticide has resulted in the identification of neem as the best alternative to toxic pesticides due to its biodegradability, relatively low toxicity and abundance. In addition, neem has been now shown to exhibit various pest control activities against as many as 300 insect species and plant pathogens (Jayaraj *et al.,* 1993).

Table 1. Important viruses and their insect vectors in vegetable crops

Crop	Virus	Vector
Chenopodiaceae-Spinach	Cucumber mosaic virus	Aphids
Brassicaceae Cauliflower	Cauliflower mosaic virus	Aphids
Cucurbitaceae Cucumber	Cucumber mosaic virus	Aphids
Pumpkin		Aphids
Watermelon	Watermelon mosaic virus I	Aphids
	Watermelon mosaic virus II	Aphids
	Tomato spotted wilt virus	Thrips
Leguminaceae:		
Beans	Bean common mosaic virus	Aphids
	Bean yellow mosaic virus	"
	Cucumber mosaic virus	"
	Bean golden mosaic virus	Whitefly
	Southern bean mosaic virus	Beetle
Cowpea	Black-eye cowpea mosaic virus	Aphids
	Cowpea aphid-borne mosaic virus	Aphids
	Cowpea banding mosaic virus	Aphids
	Cowpea golden mosaic virus	Whitefly
	Cowpea chlorotic mottle virus	"
	Southern bean mosaic virus	Beetle
	Tomato spotted wilt virus	Thrips
Cluster bean	Cluster bean mosaic virus	Aphids
	Cluster-bean yellow mosaic virus	Whitefly
Pea	Cucumber mosaic virus	Aphids
	Pea seed-borne mosaic virus	Aphids
	Tomato spotted wilt virus	Thrips
Malvaceae:		
Okra	Yellow vein mosaic virus	Whitefly
	Enation leaf curl virus	"
Solanaceae		
Brinjal	Cucumber mosaic virus	Aphids
	Alfalfa mosaic virus	"
	Potato virus Y	"
Chilli/ Capsicum	Cucumber mosaic virus	"
	Potato virus Y	"
	Pepper veinal mottle virus	"
	Tobacco etch virus	"

Crop	Virus	Vector
	Tomato spotted wilt virus	Thrips
	Tobacco leaf curl virus	Whitefly
	Tomato mosaic virus	No vector
Tomato	Cucumber mosaic virus	Aphids
	Potato virus Y	"
	Tomato leaf curl virus	Whitefly
	Tomato spotted wilt virus	Thrips
	Tomato mosaic virus	No vector

Biological and antifeedant activity: Neem tree has emerged as a single most important source of insecticides. All parts of the tree are biologically active. The maximum insecticidal activity is in seed kernel. The kernel extracts and pure compounds isolated from the seed have shown diverse biological effects against insects. These include the repellent, feeding and oviposition, deterrent, growth regulatory (IGR) and sterilant effects. In addition, neem is also reported to have direct toxicity and impairs egg hatchability (Singh, 1993a).

Antifeedants - the chemicals which when perceived, reduce or prevent insect feeding and consequently, the growth, devlopment, survival and reproduction, are important defence barriers in some plants against insect attack. Neem also has been evaluated for antifeedant activity which showed varying degree of sensitivity to various extracts and pure compounds.

Active ingredients: Azadirachtin is considered the most important active principle in neem seed Kernel (NSK). A considerable number of other active compounds have been isolated from neem seed kernels such as Salannin, Salannol, Salannolacetate, 3-deacetylsalanin Azadiradion, 1G-epoxyazaradin, gedunin, nimbinen and 6-deacetylnimbinen (Jones *et al.,* 1989). Some vilasinin derivatives with strong antifeedant activities have been isolated from neem seed oil (Kraus *et al.,* 1987). Neem leaves contain some useful compounds like nimbin, nimbinene, nimbandiol and quercetin which are useful for pest and disease control.

The bio-efficacy results have attracted the attention of the pesticide industry in India and abroad. Nearly 40 compounds are either marketed or are awaiting commercialization in India (Table-2). Most of the production are either oil-based or based on various extractives.

Plant viruses are the group of pathogens that are most dependant on an intermediary, or vector for their transfer from plant to plant. Viruses cannot be controlled in infected plants by the application of chemicals, unlike fungi and to a lesser extent bacteria. Therefore control at present depends upon preventing infection. This can be done by avoiding the vectors that carry viruses, by manipulating sowing dates or by modifying cropping patterns. Alternatively the vectors can be killed, either before they transmit the virus or by limiting the number of plants they can infect or plants can be bred resistant to either virus or vector.

Table 2. Commercially produced neem-based pesticidal products in India

Product	Active ingredients	Activity claimed	Manufacturer
1. Godrej Achook	Azadirachtin, Azadiradione, Nimbocinol Epinimbocinol	Antifeedant repellent/deterrent, male sterility, molt/chitin inhibitor, change in sexual behaviour, growth regulator/disruptant, ovipositional deterent juvenile hormone, interference in biosynthesis, foscicant, ovicidal, oviposition deterrent altered egg hatchability.	M/s. Bahar Agrochem & Feeds Pvt. Ltd. M/s. Godrej Agrovet Ltd., Bombay-400 079.
2. Field Marshal	Azadirachtin	Antifeedant/repellent	M/s. Khetiwadi Corner, Vadodara-390 001, Gujarath.
3. Jawan Crop Protector	Extracts containing Azadiracthin	Antifeedant/repellent, disturbs growth and reproduction.	M/s. McDA Agro Pvt. Ltd., Bombay-400 001.
4. Margocide-CK	Azadirachtin and its 15 derivatives	Antifeedant, growth inhibitory, ovicidal, oviposition deterrent, nematicidal, chemosterilant.	M/s. Monofix Agro-products Ltd., Hubli-580 029.
5. Margocide-OK	Azadirachtin	Insecticidal	- do -
6. Neem based Emulsifiable concentrate, Dust, water Dispensible powder, Granule	Kernel or oil	Pesticidal	IARI, New Delhi-110012.
7. Neem oil emulsion	Oil	Pesticidal	M/s. Sio Agro Research Labs, Bombay-400 080.
8. Neemark	Azadirachtin	Antifeedant/repellent, nematicidal, synergistic.	M/s. West Coast Herbochem Pvt. Ltd., Bombay-400 025.
9. Neem gold	Azadirachtin Kernel Extract	Antifeedant	M/s. Southern Petro Chemicals Corpn. Ltd., Madras-32.
10. Neem guard	Extract concentrate	Repellant, metamorphsis disruption	M/s. Gharda Chemicals Pvt. Ltd., Bombay-050.
11. Neemta 2100	Extract concentrate	Repellant, metamorphsis disruptor	M/s. A.J. Chemicals, Ahmedabad-380 002.
12. Nethrin	Oil	Pesticidal	M/s. Amitul Agrochem Pvt. Ltd., Gorakhpur-273 001, U.P.
13. Nimba	Kernel-based powder	Pesticidal	IARI, New Delhi-012.
14. *Nimbecidine	Azadirachtin	Antifeedant, repellent, metamorphsis disruptor	M/s. T. Stanes and Co. Ltd., Coimbatore-018.
15. Nimlin	Extract concentrate	Repellent, metamorphosis disruptor	M/s. Sunline Agro Chemicals, P.B. No. 73, Dhulia, Maharastra.
16. *RD-9 Repelin	Neem, Pongamia, Annona and Castor (Azadirachtin 3000 ppm)	Antifeedant, repellent Rajamundry, A.P.	M/s. ITC Ltd., ILTD,
17. Sukrina	Azadirachtin, Meliantriol	Antifeedant, repellent	M/s. Conser Chemicals Pvt. Ltd. Madras-016.
18. Wellgro	Neem Kernel Powder	Repellent, fungi inhibitor, Antiviral	M/s. ITC Ltd. ILTD, Rajamundry, A.P.
19. Suneem	Extract concentrate	Antifeedant/repellent	Sunida Exports, Bombay-400 049.

* Registered with Central Insecticides Board.

Neem pesticides as inhibitors or antiviral agents: The occurrence of inhibitory substances in plant extracts was first noticed when sap transmissible viruses could not be transmitted from certain host plants to other hosts. These inhibitory substances have been referred to as antiviral agents. When the extracts from such inhibitory plants were mixed with the infective sap of viruses, there were different degrees of virus inhibition (Verma, 1986).

The chemical characteristics of the inhibitors from various plants and the mechanism by which they reduce virus-infection, show that the problem of inhibitors present in different plant extracts is a complex one and the inhibitors present in different plants may display individuality.

Most studies on antiviral agents from plants involved mixing of inhibitor with virus before mechanical inoculation or application of inhibitor to the leaves at various times before or after inoculation with virus. These inhibitors either normally do not act at a distance from the point of application or alternatively the method used were inadequate to assess this possibility.

Indeed, rather impressive amount of advanced research has already been reported on the occurrence, identity, and mode of action of a variety of antiviral agents in plants (Gupta, 1977; Verma, 1986). There are several reports on the use of neem and its products against plant virus inhibition (Verma, 1974; Zaidi *et al.*, 1988; Singh *et al.*, 1988; Nagarajan *et al.*, 1990). Verma (1974) reported that Nimbidin and Nimbin obtained from neem oil, were active against viruses. Nimbidin at 1000 ppm inhibited potato virus X (PVX) by 62% whereas Nimbin at the same concentration was 50% inhibitory. The leaves of *Chenopodium amaranticolor* sprayed with these chemicals before or after inoculation reduced local lesion formation. Pre-inoculation root treatment of tomato seedlings at 500 ppm did not delay symptoms but when *C. amaranticolor* seedlings were treated with a mixture of both the compounds at 250 ppm or with Nimbidin at 500 ppm, the local lesions were reduced upto 65 and 35% respectively.

Inhibitory effects of the extracts of *A. indica* were found on Spinach mosaic virus on the host *Chenopodium amaranticolor* (Zaidi *et al.*, 1988). The inhibition was directly correlated with concentration of the extract. It is possible that extract might contain certain compounds which bring about inhibitory effects. Neem leaf extract when mixed with leaf extract of PVX in 1:0.5, 1:1, 1:1.5 and 1:2 ratios and inoculated to *Capsicum pendulum* complete inhibition was observed at the ratio of 1:2. Spraying the neem leaf extract on upper surface of the leaves of the test plants was found effective in completly preventing virus infection upto 4 hr. but its efficacy decreased gradually by increasing the interval between treatment and inoculation (Sangar and Dhingra, 1982). The neem leaf extract exhibited no phytotoxicity. The inhibitor in aqueous neem leaf extract is thermostable, resists storage and a wide-range of pH (Sangar and Dhingra, 1982).

Singh *et al.* (1988) worked on the antiviral activity and physical properties of the extracts of *Azadirachta indica.* The leaf and bark extracts inhibited infection of *Chenopodium amaranticolor* by cowpea mosaic comovirus. The inhibition was heat stable, non-dialysable through cellophane, remained active after 25 days storage, not destroyed by desiccation for 50 days, showed appreciable loss of activity on dilution, was destroyed by activated charcoal and induced localized resistance. Antiviral activity was lowered in strong acidic and basic solutions and it was reversed by simultaneous application of Actinomycin D. The extracts were not inhibitory when applied on virus inoculated plants but were highly effective when applied to the leaves prior to inoculation. Nagarajan *et al.* (1990) reported that aqueous extracts of neem possessed strong anti-viral substances against TMV infection inhibiting upto 86.8%.

Expressed juices of plant parts from different flowering-fruiting plants were found to contain a substance which markedly decreased the number of local lesions and systemic infection by the virus causing necrotic mosaic disease on *Solanum melongena* (Gupta and Naqvi, 1991). The extent of the antiviral activity of the substance was independent of the species from which juice was extracted as well as the concentration and method of application of expressed juice, Extremely acidic or alkaline PH of the expressed juice caused the inhibitory activity.

Control based on vector behaviour: Recent developments of insect antifeedants and their use to control virus vectors are reviewed by Griffiths *et al.* (1989). The probing and feeding behaviour of insect vectors can result in uptake of viruses from infected plants and subsequent transmission to healthy plants. It is possible to interfere with virus acquisition and transmission by influencing vector host-selection and feeding behaviour with antifeedant chemicals.

It is not surprising that the process can be disrupted by repellent or antifeedant materials because efficient transmission takes so long. For the non-persistent process, however it could be thought that virus-spread might be encouraged by making aphids disperse either by contact with an antifeedant or as a response to an alarm pheromone. That this does not appear to happen is encouraging for future studies, although the mechanism of action of such materials is as yet unknown (Griffiths *et al.,* 1989).

Antifeedant Activity: Insects evaluated with neem for antifeedant activity show varying degree of sensitivity to various extracts and pure compounds irrespective of the order or the family of the insect. Cabbage and cauliflower crops sprayed with NSKE and aqueous extract of leaf showed strong antifeedant activity against aphid *Brevicorne brassicae* (Singh and Sharma, 1986). Extracts of neem seeds had some antifeedant activity when applied to artificial membranes (Griffiths *et al.,* 1978), but failed to deter settling and feeding on plants. Pure azadirachtin, a limonoid tetranortriterpenoid known to be insecticidal and antifeedant against other insects (Schmutterer, 1988), was only weakly active against aphids.

The botanical product RD-Repellin was highly repellent to pea aphids at concentrations of 1%, 4% or 10% which transmits Zucchini yellow mosaic virus (ZYMV).

Behavioural observations revealed that this repellency included both pre and post-probing components. Aphids contacted treated leaves with lower frequency than untreated leaves and more often abandoned treated leaves and initiated wandering behaviour. Aphid behaviour suggests that aphids may detect RD-Repellin through olfactory stimuli. It is likely that the repellent and toxic effects of D-Repellin could reduce the build up of colonising aphid populations in commercial plantings. The increased tendency of aphids to initiate wandering behaviour after contact with RD-Repelin could significantly affect aphid population dynamics and, consequently, virus spread in agroecosystems. Wandering aphids cannot readily feed and may, therefore, have an increased risk of mortality due to dehydration. Wandering aphids may also increase their risk of detection and exploitation by natural enemies, including parasitoids and predators (Hunter land Ullman, 1992). Although these potential sources of increased mortality might act to suppress virus transmission, increased aphid mortality might also accelerate the spread of virus from infection sources as aphids dispuse in search of acceptable hosts.

Reduction of feeding was also observed after topical application or injection of neem derivatives including Azadirachtin and alcoholic neem seed kernel extract. This means that it is regulated by sensory organs of the mouthparts, but also non-gustatory. These two phagodeterrent/antifeedant effects were called primary and secondary effects of neem (Schmutterer, 1985).

Insect growth regulatory effect and toxicity: Growth regulator effect is the most important physiological effect of neem on insects. The seed kernel extracts or azadirachtin when fed or applied at juvenile stages arrest their growth. Depending on dose, the insects are either killed before reaching adult stage or produce malformed and miniature adults (Singh, 1993).

How azadirachtin affects insects is still not fully understood. It is, however, known that axadirachtin reduces and delays titers of morphogenetic hormones. There is also evidence that azadirachtin interferes with neuroendocrine system which controls ecdysone and juvenile hormones, but how it is accomplished is still unclear (Singh, 1993a).

There are several reports which suggest that neem has direct toxicity. Nonfatty alcoholic extract of the seed cake was reported to possess good contact toxicity against mustard aphid *Lipaphis erysimi* (Sinha, 1960). Alcohol extract of cake has also been reported to be toxic against another aphid, *Rhopalosiphum nympheae* (Goyal, 1971).

Neem and lemon grass products have been tested on various aphid vectors (*Aphis craccivora, A. gossypii, A. rumicis* and *Myzus persicae*) of plant viruses. It was found that neem leaf extract, seed kernel extract and neem oil treatment of plants influenced the nymphal development time and ecdysis of most aphids and was also toxic to them. Commercial neem soap and detergent solutions were also equally toxic. Neem oil (2-4%) and neem soap sprays on *A. rumicis* caused almost complete death of the nymphs and alate forms within 24h while anapterous adults died in two days. Neem oil and soap solutions were more effective on *A. craccivora* than Citronella oil (Chaudhury and Jain, 1989). Toxicity of neem oil may be due to ingestion or contact poison (Parveen and Alam, 1993).

Oviposition deterrent effect: Oviposition deterrency is an important behavioural effect of neem on insects and is very useful in protecting crop especially in integrated pest management. Both neem seed oil and water extract of neem seed kernel have been reported to deter oviposition of several insects.

Neem seed extract, containing a tetranortriterpenoid, isolated from the seeds of the neem was used for evaluation against immature stages of *Bemisia tabaci.* Cotton foliage treated with neem-seed extract resulted in a reduction of egg viability and oviposition, prolonged larva periods, and larva mortality (Coudriet *et al.,* 1985). Although the mode of action of the extract is known, it was observed that adults were repelled, thus reducing oviposition. The failure of treated larvae to complete ecdysis would indicate an antiecdysteroid activity.

Pea aphid, *A. pisi* which were kept from the first nymphal instar onwards on *Vicia faba* plants treated with 20 ppm per litre of purified neem seed kernel extract, produced only about one twelth the number of nymphs of insects in control (Schauer, 1985).

Effect on vector transmission of plant viruses: The control of viruses carried passively can be achieved only by preventing contact between vector and virus. Viruses infecting plants cannot be suppressed by chemicals, therefore control has to be directed at the vector. The biological interactions between vector and virus offer opportunities for interfering with the transmission processes without recourse to toxic material.

Srivastava *et al.* (1986) found that crude margosa oil (0.5% water emulsion) inhibited the transmission rate of cucumber mosaic virus (CMV) by single apterous aphid to the extent of 40% when the aphids were allowed to come in contact with oil before acquisition, 33.3% during acquisition 23.3% between acquisition and inoculation and 30% inhibition at the time of inoculation respectively. Further no effect was found on transmission, when plants were sprayed with oil emulsion after inoculation. However, oil did not affect the biological activity of the virus as evidenced by a mechanical transmission test using a mixture of oil and virus. A definite influence of the oil on the feeding behaviour of the aphid transmission of CMV by margosa oil is either related to change in feeding behaviour of aphids as induced by other mineral oils (Simons *et al.* 1977) or due to electric insulating properties of the oil which would hamper the exchange of charges between virus particles, aphid mouth parts and plant cells as suggested by Zschiengner *et al.* (1974).

Dube and Nene (1974) used margosa oil against cowpea mosaic virus and obtained about 50% inhibition in aphid transmission with 2% oil emulsion.

Singh (1993) reported that two neem products i.e. Repelin and Neemark significantly inhibited the acquisition and inoculation of tomato leaf-curl virus by the whitefly vector *Bemisia tabaci.* Repellin at 2% and Neemark at 2.5% concentration were most effective. Roychoudhury and Jain (1993) reported neem oil inhibition of okra yellow vein mosaic transmission by *Bemisia tabaci.*

Non-edible seed oils of neem and mahua were tested individually and in mixtures for their effect on the insect vectors *Aphis craccivora* Koch. and *Myzus persicae* Sulz. The seed oil preparations (3% and 5%) were not found to impede the transmission of PVY effectively when inoculated mechanically. However, the transmission of PVY to chillis by aphids was impeded to a considerable extent. Among the oils, neem oil and mahua oil when applied individually were more effective than when applied in mixtures; neem oil was more effective (Soundar Rajan and Nambisan, 1993).

Induced resistance to viruses: Induced resistance to virus diseases occurs when, as a result of a primary inoculation or chemical treatment, a plant becomes resistant to a second or challenge inoculation by a related or unrelated pathogen (Sela and Applebaum, 1962; Loebenstein and Ross, 1963; Nagaich and Singh, 1970). The resistance induced may be localized or systemic. Such phenomena are also frequently referred to by the terms 'acquired systemic resistance'. The word 'induced' is preferred here purely because it implies some active process, rather than the passiveness which 'acquisition' tends to imply.

Active defence mechanism against viruses exists in plants and it can be stimulated by antiviral agents identified in a few plants species. Activation of such mechanisms results in plants becoming highly resistant to subsequent infection. Molecular basis of such mechanism may involve synthesis of new proteins in which host DNA plays an important role. The active defence mechanisms in plants against viruses are not completly different from interferon system in animals. The plants extracts activated a resistance mechanism in the treated plant tissues and the resistance is also induced in non-treated tissues after other parts of plants have been treated with the plant extracts. They induce anti-viral state by acting through actinomycin sensitive mechanism, mostly involving synthesis of new virus inhibitory substances by the host. Neem extracts were also known to have such properties of inducing resistance to viruses.

Aqueous flower extract of *Azadirachta indica, Euphorbia milii,* induced resistance against potato virus X infection in hypersensitive host (*Chenopodium amaranticolor*). The resistance in the host varied from 79.2 to 99.2 percent and remained active upto 96 h after their application as spray on the upper surface of the host leaves (Rao *et al.,* 1985). In nonhypersensitive host, tomato (*Lycopersicum esculentum* Mill.) the aforesaid flower extract induced varying degree of systemic resistance against PVX infection.

CONCLUSIONS

Utility of neem and its derivatives in pest control has been established beyond any doubt. Investigations with respect to neem bioactivity against plant viruses are limited and of preliminary in nature. However, neem possesses some anti-viral components which is highly desirable to further explore and then exploit these components for practical control of plant viruses and their insect vectors. Owing to their mode of action neem derivatives could be very suitable for integrated virus disease management. They are primarily feeder poisons for nymphs/larvae of insect vectors of plants and therefore, they show a considerable selectivity toward natural enemies of pests, especially parasites and numerous predators.

The relatively short residual life of active principles of neem derivatives may be considered a disadvantage from a purely economic point of view, especially in cases of a long-lasting pest pressure. From the ecological stand-point, however, it should be kept in mind that products with such properties will not disturb ecosystems and consequently will not cause outbreaks of new pests, as long-lasting insecticides are apt to do. Resource poor farmers in developing countries can grow their own neem trees and produce simple toxic pesticides for crop pest control. Neem products can be mixed with other biproducts and farm practices.

The unique mode of action of neem derivative interferring with the hormonal system of insects, especially ecdysteroids may open new ways of environmentally sound pest control by the use of botanicals. Biodegradability of neem products certainly gives an edge over the conventional pesticides where there are problems of residual toxicity and pollution hazards. The occurrence of a pod of biologically active and degradable allelopathic chemicals probably holds the key for future exploitation of neem in plant virus control.

Research into the molecular basis of viruses and their vectors interactions is just beginning. There is good evidence of the vital role played by proteins in these interactions between nematodes, whiteflies, fungi and leafhoppers, as well as that between aphids and viruses. As crop protection practices more durable and environmentally acceptable approached to pest and disease control knowledge at the molecular level of plant/virus/vector interactions will play an important role. This seems likely to be principally through the medium of plants genetically transformed to combat or interfere with a particular virus/vector interaction and may dispense with the need for any externally applied crop protection agents.

REFERENCES

Chaudhury, R.R. and R.K. Jain. 1989. Effect of neem and lemongrass products on some aphid vectors of plant viruses. *Indian Phytopath.*, **42**: 230.

Coudriet, D.L., N. Prabhakar, and D.E. Meyerdirk. 1985. Sweet potato whitefly (Homoptera: Aleyrodidae): Effects of Neem-seed extract on oviposition and Immature stages. *Environ. Entomol.,* **14**: 776-779.

Dube, G.S. and Y.L. Nene, 1974. Aphid transmission of cowpea mosaic virus as influenced by oil sprays. *Indian Phytopath.*, **27**: 325-330.

Goyal, R.S. 1971. Studies on the better utilization of neem (*Azadirachta indica*) seed cake. Biological activity of alcohol extract and its isolates. M.Sc. thesis, Post-graduate school, IARI, New Delhi.

Griffiths, D.C., A.R. Greenway and S.L. Lloyd. 1978. The influence of repellent materials and aphid extracts on settling behaviour and larviposition of *Myzus persicae* (Sulzer) (Hemiptera: Aphididae). *Bull. Entomol. Res.*, **68**: 613-619.

Griffiths, D.C., J.A. Pickett, L.E. Smart and Woodcock, C.M. 1989. Use of insect Antifeedants against aphid vectors of plant virus disease. *Pestic. Sci.*, **27**: 269-276.

Gupta, B.M. 1977. Inhibition of plant virus infections by antiviral agents. *In*: Aphids as virus vectors (eds.) K.F. Harris and K. Maramoto Sch. Academic Press, New York. pp. 455-471.

Gupta, V.P. and Q.A. Naqvi. 1991. Inhibitory activities of Plant extracts on virus infection. *Acta. Bot. Indica*, **19**: 18-24.

Hunter, W.B. and D.E. Ullman. 1992. Effect of the neem product, RD-Repelin on settling behaviour and transmission of Zucchini yellow mosaic virus by the pea aphid, *Acyrthosiphon pisum* (Harris) (Homoptera: Aphididae). *Ann. Appl. Biol.*, **120**: 9-15.

Jayaraj, S., M. Bharathi and P.C. Sundara Babu. 1993. Integrated pest management. *In*: Neem-Research and Development (eds.) Randhawa and Parmar. Society of Pesticide Science, India.

Jones, P.S., S.V. Ley, E.D. Morgan and D. Santafianos. 1989. The chemistry of the neem-tree. *In*: 1988 Focus on phytochemistel pesticides, Vol. I, The neem Tree, (Ed) Jacobson, M. Boca. Raton. CRC Press, pp. 19-45.

Kraus, W., S. Baumann, M. Bolcel, U. Keller, A. Klenk, M. Klingele, H. Pohul and M. Schwinger. 1987. Control of insect feeding and development by constituents of *Melia azedarach* and *Azadirachta indica*. *In*: natural pesticides from the neem tree and other Tropical plants, Proc. 3rd Int. Neem Conf. (eds.) H. Schmutterer, K.R.S. Ascher, Nairobi, Eschborn, GTZ, pp. 111-125.

Loebenstein, G. and A.F. Ross. 1963. An extractable agent induced in uninfected tissue by localized virus infection, that interferes with infection by tobacco mosaic virus. *Virology*, **20**: 507-517.

Nagaich, B.B. and S.J. Singh. 1970. An antiviral principal induced by potato virus X inoculation in *Capsicum pendulum* Willd. *Virology*, **40**: 267-271.

Nagarajan, K., N.S. Murthy and T.S.N. Reddy. 1990. Utilization of botanicals possessing anti-viral principle against Tobacco mosaic virus. Natl. Symp. on Problems and prospects of botanical pesticides in Integrated Pest management, CTRI, Rajmundry, 21-22, January, pp. 39. (Abst.).

Parveen, G. and M.M. Alam. 1993. Bio-activity against plant pathogens. *In*: 'Neem Research and Development, (Eds.) N.S. Randhawa and B.S. Parmar, Publication No. 3, Society of Pesticides Science, India, pp. 144-153.

Rao, G.P., R.K. Singh, A.K. Pandey and R. Singh. 1985. Induction of local and systemic resistance of potato virus X by flower extracts. *Indian J. Virology*, **1**: 174-182.

Roychoudhury, R. and R.K. Jain, 1993. Neem for the control of aphid and whitefly vectors and virus diseases of plants. *In*: World Neem Conference, pp. 31. (Abst.).

Sangar, R.B.S. and M.K. Dhingra. 1982. Potato virus X inhibitor from neem leaf extract. *J. Indian Potato Assoc.*, **9**: 148-149.

Schauer, M. 1985. Die Wirkung Van Nieminhaltsstoffen auf Blattlause Und die Rubenblattwanze. Ph.D. Thesis, University Giessen, FRG.

Schmutterer, H. 1985. Which insect pests can be controlled by application of neem seed kernel extracts under field conditions? *Z. Angew. Entomol.*, **100**: 468-475.

Schmutterer, H. 1988. Potential of azadirachtin-containing pesticides for integrated pest control in developing and industralized countries. *J. Insect Physiol.*, **34**: 713-719.

Sela, I. and S.W. Applebaum. 1962. Occurrence of antiviral factor in virus-infected plants. *Virology*, **17**: 543-548.

Simons, J.N., D.L. Mclean and M.G. Kinbey. 1977. Effect of mineral oil on probing behaviour and transmission of stylet borne viruses by *M. persicae*. *J. Econ. Ent.*, **70**: 309-315.

Singh, A.K., M. Singh and A.K. Singh. 1988. Antiviral activity and physical properties of the extracts of *Azadirachta indica* L. *Indian J. Virol.*, **4**: 76-81.

Singh, K. and U.L. Sharma. 1986. Studies on antifeedant and repellent qualities of neem (*Azadirachta indica*) against aphid (*Brevicorne brassicae*) on cauliflower and cabbage. *Res. Dev. Reptr.*, **3**: 33-35.

Singh, R.P. 1993. Bio-activity against insect pests. *In*: Neem Research and development, (Eds.) N.S. Randhawa and B.S. Parmar. Publication No. 3, pp. 109-122.

Singh, S.J. 1993. Effect of repelin and neemark on the acquisition and transmission of leaf-curl virus of tomato through whitefly (*Bemisia tabaci* Gen.). *In*: World Neem Conferenc. pp. 42. (Abst.).

Sinha, N.P. 1960. Studies on the better utilization of neem (*Azadirachta indica*) seed cake. Biological activity of alcohol extract and its isolates. M.Sc. Thesis, Post-graduate School, IARI, New Delhi.

Srivastava, K.M., N.S. Rana, V.C. Dwadash Shreni, and B.P. Singh. 1986. Mechanism of inhibition of CMV by Crude oil from Margosa (*Azadirachta indica*) during inoculation with single apterous *Aphis gossypii*. *Indian Phytopath.,* **39**: 20-25.

Soundar Rajan, K. and K.M.P. Nambisan. 1993. Effectiveness of neem against five temperate crop pests in the upper palani hills. *In*: World Neem Conference, pp. 28. (Abst.).

Verma, H.N. 1986. Interferon like antiviral agents from plants. *In*: Vistas in plant pathology (eds.) A. Varma and J.P. Varma. Malhotra Publishing House, New Delhi. India. pp. 481-490.

Verma, V.S. 1974. Chemical compounds from *Azadirachta indica* as inhibitors of potato virus X. *Acta Microbiol. Pol. B.* **6**: 9-13.

Zaidi, Z.B., V.P. Gupta, A. Samad and Q.A. Naqvi. 1988. Inhibition of Spinach mosaic by extracts of some medicinal plants. *Curr. Sci.,* **57**: 151-152.

Zschiengner, H.J., K. Eiesenbrandt and R. Fritzsche. 1974. Beitreg Zum Wirkungs mechanisms Van Mineralolen bei der Verhinderung der virus bertragung durch vektoren. *Arch. Phytopathol. U. Pflanzenschutz. Berlin,* **10**: 371-382.

Neem for the Management of Crop Diseases, 1995
Editor : V. Mariappan
Associated Publishing Co., New Delhi, India
pp. 137-144.

USEFULNESS OF NEEM PRODUCTS FOR THE MANAGEMENT OF VIRUS DISEASES OF URDBEAN, MUNG BEAN AND SOYBEAN

Sethuraman, K., Mariappan, V. and Yesuraja, I.

Department of Plant Pathology, Tamil Nadu Agricultural University
Coimbatore-641 003, Tamil Nadu

Urbean and mungbean are the important pulse crops grown in Tamil Nadu both as irrigated crop as well as dry crop. These crops are also grown in rice fallows utilizing the residual moisture after the harvest of rice crops. In recent year soybean has been introduced in Tamil Nadu and grown in large areas. These crops are invariably affected by diseases caused by viruses. Management of virus diseases by controlling the insect vectors has been the usual practice among farmers. But invariably the use of insecticides have not yielded the satisfactory results since insects could transmit the disease before they were killed. Neem products possessing insecticidal and antifeedant properties have been reported to be more effective in the management of virus diseases. Hence the neem products were tested and their observations are presented in this paper.

Urdbean (Vigna L. Hepper): Urbean is an important pulse crop, very rich in protein (23.9 per cent) and carbohydrate (60.4 per cent) and is an important source of protein in Indian diet. The crop is grown in more than 1,98,000 ha in Tamil Nadu with an annual production of 52000 tonnes, accounting to an average production of 263 kg/ha (Anon, 1987).

The crop is affected by many diseases and in recent years urdbean leaf crinkle virus (ULCV) has been reported from several states of India. The symptom consist the enlargement of leaflets followed by crinckled surface of lamina in the young leaves and crinckling and rugosity of older leaves (Nene, 1972) and the symptom appear usually 3-4 weeks after sowing. As a result the plant looks bushy in appearance. The flowering was reduced and the buds contianed sterile pollen (Kolte and Nene, 1979). The virus is reported to be transmitted through seeds and also by sap (kolte and Nene, 1972). Transmission of the virus by many insect species, *Bemisia tabaci* (Narayanasamy and Jeganathan, 1973). *Aphis cracivora* and *A. gossypii* (Dhingra, 1975), leaf hopper *Circulifer tenellus* (Khatri *et al.*, 1971) beetle *Hemosepilachne dodecastigma* (Beniwal and Bharathan, 1980) aphid *Acrythosiphon pisum* (Dubey *et al.*, 1985). Most of the varieties are susceptible to the virus. A loss of 62 per cent has been reported by Kolte and Nene (1979). Ravinder Reddy (1987) tested the efficacy of plant extracts on the transmission of urdbean leaf crinckle virus and found some of the plant extracts (*Mirablis jalapa, Catharanthus roseus, Datura metal, Brugainvillea spectablis, Boerhoevia diffusa and Azadirachta indica* were effective in reducing the transmission of ULCV to 30 to 50 per cent as against 80 per cent transmission in control (Table 1). Ravinder Reddy, (1988)

Table 1. Effect of plant extracts on ULCV transmission

Plant species	ULCV incidence (%)
Mirabilis jalapa	30
Catheranthus roseus	40
Datura metal	50
Bougainvilla spectablis	50
Boerhavia diffusa	50
Azadirachta indica	50
Control	80
CD	1.41

Ravinder Reddy, 1988

tested the efficacy of DHT (0.1 per cent) fresh butter milk (10 per cent) leaf extracts of *Mirabilis jalapa* (10 per cent) Neem oil (1 per cent with teepol 0.1 per cent). A quantity of 500 lit/ha was used and two sprayings at 7 and 22 DAS were given. The per cent diseased plants were scored and presented in the table 2.

Table 2. Effect of chemicals, butter milk, plant extract and neem oil on ULCV disease spread

Treatments	Disease incidence (%)					
	28	35	42	49	56	Mean
DHT	0.09	0.89	1.07	5.89	9.01	3.39
Thuja	0.27	1.07	1.97	6.52	11.16	4.19
M. Jalapa	0.18	0.81	1.79	6.43	10.81	4.01
Butter milk	0.09	0.72	1.43	6.43	10.45	3.85
Neem oil	0.26	0.81	1.34	5.09	11.51	3.51
Control	1.07	1.63	2.86	9.02	14.47	5.85
Mean	0.33	0.99	1.75	6.59	11.24	
CD (P = 0.05)		Days	-	0.49		

Treatments - 0.54
Ravinder Reddy, 1988

Among the treatments DHT scored significantly less ULCV disease followed by other treatments including neem oil at all periods of observation. The rest of the other treatments were also effective in reducing the disease.

During 1992, a field trial was conducted to compare the efficacy of neem oil (NO 3.0 per cent) neem seed kernal extract (NSKE 5.0 per cent) and monocrotophos (1000 ml/ha) in reducing the ULCV. The results are presented in table 3.

Table 3. Effect efficacy of neem products in management of disease of urdbean (Chemmedu)

Treatments	ULCV incidence (%)	Yield (Kg/ha)	Yield Value (Rs.)	Treatment cost (Rs.)	Net gain (Rs.)
NSKE 5 per cent	18.3^{a}	842^{a}	7060	440	6620
NO 3 per cent	18.7^{a}	853^{a}	6824	365	6459
Monocrotophos (0.1%)	16.8^{a}	901^{a}	7212	235	6977
Control	25.6^{b}	715^{b}	5720	—	5720
CD	2.7	NS			

Variety - T9 March 92 to June 92
NSKE - Neem Seed Kernel Extract
NO - Neem oil

The neem preparations were as effective as insecticide in reducing the incidence of ULCV and also increased the yield significantly. Another field trail conducted at Mangalakudi, Madurai revealed that the neem products were as effective as insecticide in reducing the incidence of yellow mosaic and leaf crinkle virus diseases of urd bean. Among the neem products neem seed kernel extract has been more effective in reducing the disease and also in increasing the yield on par with the insecticide (Table 4).

Table 4. Efficacy of neem products in the control of diseases of urdbean, TNAU, India

Treatment	Mangalakudi (Madurai)[a]		
	Yellow mosaic (%)	Leaf crinkle (%)	Yield (kg/ha)
NSKE	38.3^{b}	16.4^{b}	1538^{a}
NO	27.0^{a}	13.4^{a}	1215^{b}
Monocrotophos	29.0^{a}	10.5^{a}	1755^{a}
Farmer's practice	35.3^{b}	14.7^{b}	1425^{ab}
Control	64.6^{c}	24.7^{c}	743^{c}
CD	3.0	3.7	281

[a]In a column, means followed by the same letter are not significantly different by DMRT.

Mungbean - *Vigna radiata* [(L) Wilezek)] : The mungbean is affected by several insect pest and disease. Among them yellow mosaic virus disease is of significant importance since it causes great losses in tropical and sub tropical Asian countries like India, Pakistan, Bangladesh, Srilanka, (Shukla, 1979), Thailand (Thongmeerkom *et al.*, 1981) and New Guinea (Pearson, 1981). The disease is characterised by the presence of bright yellow patches on the leaves alternating with green patches. In severe infection yellow patches are seen on pods also (Nene, 1968). Depending upon the severity the loss in yield ranging from 30-70 per cent (Marimuthu *et at.*, 1981) in Tamil Nadu 40-67 per cent in Hariyana (Singh, 1980) and 88.5 per cent in Delhi (Chenulu *et al.* 1979) have been reported. Transmission of the disease through seed (upto 1 per cent) has been reported by Vasudava (1958).

Transmission of sap also was reported from Thailand (Honda *et al.* 1983) but was not successful in India (Nene, 1972). The virus is transmitted by white fly [(*Bemisia tabaci* (Nariani, 1960; Sellammal and Chelliah, 1977)] and the insect remained viruliferous for a period of 4 days. Thus the insect has been mainly responsible for the spread of the disease in field.

Verma *et al.* (1985) reported that spraying leaf extracts of *Clerodenderon fragrans*, H.K., *Aerva sanquina lenta* Forsk. and root extracts of *Boerhavia diffusa* to reduce yellow mosaic incidence in mungbean under field conditions. *C. fragrans* recorded a reduction in the yellow mosaic virus infection upto 61.0 per cent as against 26 and 20 per cent infection by the other plant extracts respectively. These three treatments also increased the yield by 42.5, 21.6 and 6.6 per cent over control. Singh and Varma (1977) reported that three weekly sprays of vegetable oils and mineral oils at 1.5 per cent concentration reduced the yellow mosaic virus incidence in mungbean. Ravindra Babu (1987) tested the efficacy of seed oils of neem *(Azadirachta indica)*, Pungam (*Pungamia pinnata*) and cotton (*Gossypium* sp.) at 0.1 and 0.5 per cent concentration in reducing the yellow mosaic of mungbean in field. The results are presented in the table 5. The seed oils at both concentrations were effective in reducing the yellow mosaic virus and also decreased the disease significantly and they were on par with each other. Except cotton seed oil 0.5 per cent the seed oil treatments also enhanced the yield over control.

Table 5. Effect of oil sprays on mungbean yellow mosaic incidence and yield

Treatments	Disease incidence (%) DAS		Yield
	29	36	(kg/ha)
Neem oil 0.1%	15.22[a]	23.06[a]	567[a]
Neem oil 0.5%	14.91[a]	22.38[a]	558[a]
Pungam oil 0.1%	16.97[a]	25.12[a]	558[a]
Pungam oil 0.5%	16.15[a]	23.04[a]	566[a]
Cotton oil 0.1%	16.67[a]	24.73[a]	545[a]
Cotton oil 0.5%	16.54[a]	25.14[a]	457[b]
Control	21.52[b]	34.81[b]	403[b]

Ravindra Babu, 1987

During 1992 a field trial was conducted with Greengram (variety Co4) for the control of yellow mosaic virus disease. Neem products like neem seed kernel extract (5 per cent) neem oil (3 per cent) and monocrotophos (500 ml/ha) were applied immediately after noticing the incidence of yellow mosaic virus disease and repeated twice at 15 days interval. Neem oil (3 per cent) + monocrotophos 0.1 per cent reduced the yellow mosaic incidence to 11.05 per cent as against 28.4 per cent in control. NSKE 5 per cent was found to be superior in recording the highest yield of 961 kg/ha (Table 6).

Soybean: Soybean, *Glycine max* (L.) Merrill is an important pulse cum oil seed crop with great yield potential. It occupies an area of 52.685 million ha in the world

Table 6. Effect of neem products on the incidence of yellow mosaic virus disease of mungbean

Treatments	Yellow mosaic (%)	Yield (kg/ha)
NSKE 5 per cent	20.47[a]	961[a]
NO 3 per cent	18.58[a]	942[a]
Monocrotophos 1 ml/lit NSKE	17.46[a]	895[a]
2.5% + Monocrotophos 1 ml/lit	13.09[b]	900[a]
Neem oil 0.1% + Monocrotophos	11.05[b]	775[b]
Control	28.24	701[c]
CD	2.79	62.85

Variety - Co_4 March '92 to June '92

with a production of 100.199 million tonnes emounting to 1902 kg/ha. In India, the area under soybean is 1.73 million ha with a production of 1.5 million is affected by a number of virus diseases, among which the yellow mosaic virus is the major disease. Yellow mosaic causes considerable yield loss upto 86.7 per cent depending upon the season and variety. In India the disease was seen in Punjab (Sing *et al.,* 1971, Rataul and Singh, 1976) and Uttarpradesh (Suteri, 1974; Singh *et al.*, 1983 a,b). Suteri (1974) found that yellow mosaic in soybean was transmitted by the whitefly *Bemesia tabaci*. Ragupathy (1989) reported that Neem oil spray reduced the yellow mosaic virus incidence and increased the yield (Table 7).

Table 7. Effect of plant products, butter milk and chemicals on soybean yellow mosaic virus infection and yield

Treatments	SYMV incidence (%)	Grain yield (kg/ha)	*B. tabaci* population	
			Before spray	24 hr after last spray
	Neem oil	1.65	566.5	1.8.0
Butter milk	3.83	550.0	8.3	8.1
Sorghum bicolor	12.22	490.0	7.6	8.8
Prosopis chilensis	5.47	550.0	8.2	8.3
Acetyl salicylic acid	8.34	475.0	8.5	7.6
Methyl dematon	2.78	558.5	8.4	1.5
Monocrotophos	1.28	558.5	8.7	1.2
Control	1.28	438.5	8.5	8.9
CD	2.140			

Ragupathy, 1989

A field trial on soybean was conducted in Research Station, TNAU, Coimbatore. The NSKE and monocrotophos treated plots recorded 11.0 and 12.6 per cent yellow mosaic incidence and they were on par followed by neem oil which recorded 13.11 per cent whereas control plots recorded 20.4 per cent incidence with regard to yield NSKE treated plot recorded 1038.7 kg/ha followed by monocrotophos and neem oil (993.7 and 978.7 kg/ha) respectively whereas control plots yield only 656.2 kg/ha (Table 8).

Table 8. Field efficacy of neem products of soybean yellow mosaic disease (Coimbatore)

Treatments	Yellow mosaic (%)	Yield (kg/ha)	Yield value (Rs.)	Treatment cost (Rs.)	Net gain (Rs.)
NSKE	12.6[ab]	1038.75[a]	9342	440	8902
NO	13.1[b]	978.75[a]	8811	365	8446
Monocrotophos	11.0[a]	993.75[a]	8946	235	8711
Control	20.4[c]	656.25[a]	5904	—	5904

Variety - Co_1 March '92 to June '92
NSKE - Neem seed kernal Extract
NO - Neem oil

One of the methods for controlling insect transmitted virus disease of crop plants is the application of insecticides, that would kill the insect vector. But none of the insecticide is effective and kill the insects instantaneously before they could acquire or transmit the virus. Harrison (1984) opined that it was more difficult to decrease the number and activity of insect vectors to a point at which virus spread is effectively controlled. The usefulness of mineral oils in reducing the transmission of aphid transmitted viruses have been reported (Vandervekan, 1977). Oils have been reported to reduce the spread of certain persistent viruses by directly interfering with vector behaviour. Nene (1973) reported immobilisation of whiteflies by orchard oil and ovicidal action which checked the white fly multiplication. Increase in yield and reduction of disease were reported with 'Krishi oil' on fresh beans against yellow mosaic virus (Sastry, 1981) and on tomato against tomato leaf curl virus which are transmitted by white flies (Singh *et al.*, 1973; Sastry *et al.*, 1978). The spread of mungbean yellow mosaic virus was reported to be reduced by application of mineral oils and vegetable oils (Singh and Varma, 1977; Sharma and Varma, 1982).

The efficacy of seed oils in reducing the transmission of plant viruses that are transmitted by insect vectors have been demonstrated in recent years. The transmission of chilli mosaic virus to chilli plants that were treated with neem oil (*A. indica*) and mahua oil (*Madhuca longifolia*) at 5 per cent concentration three hours prior to the inoculation feeding by the aphid vectors *Aphis craccivora* and *Myzus persicae* were reported to be reduced to an extent of 53.3 to 73.3 per cent over control (Duleep Kumar Samuel, 1990). Mariappan *et al.* (1988) reported that the non-edible seed oil extracted from seeds of Karanja (*Pongamia pinnata*) Mahua (*Madhuca longifolia*) and punnai (*Calophyllum inophyllum*) were as effective as neem (*A. indica*) in reducing the transmission of rice tungro virus by the green leaf hopper (*Nephotettix virescens*).

Neem oil has been reported to possess some adverse effect on some vectors. Coudriet *et al.* (1985) reported that neem seed extract repelled whitefly from alighting on cotton.

These observations have clearly indicated that the neem preparations were effective in reducing the virus diseases of crop plants that are transmitted by insect vectors. The results presented in the tables have clearly indicated that the neem preparations were as effective as insecticide in reducing urdbean leaf crinkle, mungbean yellow mosaic and soybean yellow mosaic virus diseases and also in increasing the yield.

REFERENCES

Anonymous, 1987. Proceedings of Pulse Workshop, Tamil Nadu Agricultural University, Coimbatore.

Beniwal, S.P.S. and N. Bharathan. 1980. Beetle transmission of urdbean leaf crinkcle virus. *Indian Phytopath.*, **33**: 600-601.

Chenulu, V.V., V. Venkateswaralu, R. Rangaraju. 1979. Studies on yellow mosaic disease of mungbean. *Indian Phytopath.*, **32**: 230-235.

Coudriet, D.L., Prabhakar, N. and Mayerdirk, D.E. 1985. Sweet to potato whitefly (Homoptera: Aleyrodidae) Effects of neem seed extract on oviposition and immature stages. *Environ Ent.* **14**: 776-779.

Dhingra, K.L. 1975. Transmission of urdbean leaf crinckle virus by two aphid species. *Indian Phytopathology* **28**: 80-82.

Dubey, G.S. and Indusharma. 1985. Seed transmission of urdbean leaf crinkle virus. *Seed Sci. and Technol.*, **13**: 149-153.

Dubey, G.S., Indusharma and Nam Prakash. 1983. Some properties of urdbean leaf crinkle virus. *Indian Phytopathology.* **36**: 762-764.

Duleepkumar Samuel, L. 1990. Studies on chilli mosaic virus. M.Sc.(Ag.) Thesis submitted to the Department of Plant Pathology, Agricultural College and REs. Institute, Madurai, pp. 147.

FAO. 1986. Production - Soybean. *Month. Bull. of Stat.*, **9** (1): 18.

Harrison, B.D. 1984. Progress and problems in the control of Arthropod-Nematode and seed borne plant viruses. In E. Kurstak (ed.) control of virus diseases Marcel Dekkam, New York, pp. 265-289.

Honda, Y., M., Iwaki, Y., Saito, P., Thongmeerkom, K. Kittisak and N. Deema. 1983. Mechanical transmission, purification and some properties of whitefly borne mung bean yellow mosaic virus in Thailand. *Plant Dis.*, **67**: 801-804.

Khatri, H.L., Bhatia D.S. and Chohan, J.S. 1971. Brief account of the work done on diseases of kharif pulse crops at the Dept. of Botany and Plant Pathology, PAU, Ludhiana during 1970-71. Fifth All India Pulse Improvement Programme Workshop, Hissar.

Kolte, S.J. and Nene, Y.L. 1972. Studies on soymptoms and mode of transmission of the leaf crinckle virus of urd bean (*Phaseolus mungo* L.). *Indian Phytopath.*, **28**: 430-431.

Kolte, S.J. and Y.L. Nene. 1979. Urdbean (*Vigna mungo,* leaf crinckle virus: Note worthy symptoms on host and influence of growth stages on host succeptibility. *Tropical Grain Legume. Bulletin,* **15**: 5-8.

Mariappan, V. S. Jayaraj and R.C. Saxena, 1988. Effect of non edible seed oils on survival of *Nephotettix virescens* (Homoptera: Cicadellidae) and on transmission of rice tungro virus. *J. Econ. Entomol.*, **81**: (5) 1369-1372.

Marimuthu, T., C.L. Subramanian and R. Mohan. 1981. Assessment of yield losses due to yellow mosaic infection in mungbean. *Pulse crops Newsl.*, **1**: 104.

Narayanasamy, P. and Jaganathan, T. 1973. Sources of resistance to blackgram virus diseases. *Madras Agric. J.*, **60**: 1836-1838.

Nene, Y.L. 1968. First annual Report, FG-IN-358, *G.B. Pant Univ. Agric.* and *Tech.*, Pantnagar, pp. 1-25.

Nene, Y.L. 1972. A survey of viral diseases of pulse crops in Uttarpradesh. Final Tropical Report. G.B. Pant University of Agriculture and Technology, Pantnagar U.P. India. pp. 1-91.

Nene, Y.L. 1972. A survey of viral diseases of pulse crops in Uttar Pradesh. G.B. Pant Univ. Agric. Tech. Pant Nagar, Res. Bull. 4, 199 pp.

Nene, Y.L. 1973. Control of *Bemisia tabaci* Genn., a vector f several plant viruses. *Indian J. Agric. Sci.* **43**: 433-436.

Nariani, T.K. 1960. yellow mosaic of mungbean. (*Phaseolus aureus* L.). *Indian Phytopath.*, **13**: 24-29.

Pearson, M.M. 1981. A mosaic disease of mungbean in Papua. New Guinea. *Australian Pl. Path.* **10**: 31.

Ragupathy, N. 1989. Studies of yellow mosaic virus disease of soybean (*Glycine max* L.) (Merril), M.Sc.(Ag.) Thesis. Dept. of Plant Pathology, TNAU, Coimbatore-3.

Rataul, H.S. and Singh, L. 1976. Field studies on the control of soybean yellow mosaic virus, transmitted by whitefly *Bemesia tabaci* (Genno) by using different granular systemic insecticides. *J. Res.,* **13**: 298-302.

Ravinder Reddy, Ch. 1988. Studies on urdbean (*Vigna mungo* (L.) Hepper) leaf crinckle virus. Ph.D. thesis submitted to the Dept. of Pl. Pathology, TNAU, Coimbatore. 168 pp.

Ravindra Babu, R. 1987. Studies on mungbean yellow mosaic virus disease. Ph.D. Thesis. Dept. of Plant Pathology, TNAU, Coimbatore-3.

Sastry, K.S. 1981. Management of tomato leaf curl virus by Carbofuran and oil treatment. In Third Interstat. Symposium on Plant Pathology, New Delhi, p. 225.

Sastry, S.S.M., Singh, S.J. and Sastry, K.S. 1978. Efficacy of Krishi oil in relation to the control of whitefly population and the spread of tomato leaf curl virus. *Pesticides* **12**(1): 28-29.

Sellammal, M. and S. Chelliah. 1977. Transmission of greengram yellow mosaic virus by the whitefly *Bemisia tabaci* (Genn.) *Madras Agric. J.,* **64**: 437-441.

Sharma, S.R. and Varma, A. 1982. Control of yellow mosaic of mungbean through insecticides and oils. *J. Ent. Res.* **6**: 130-136.

Shukla, G.P. 1979. Ph.D. thesis. GB Pant Univ. Agric. and Tech. Pantnagar (cited after G.P. Shukla, 1986).

Singh, B.R., M. Singh and M.D. Yadav. 1983a. Estimation of yield losses in soybean due to yellow mosaic *Madras Agric. J.,* **70**: 312-315.

Singh, S.J., Sastry, K.S.M. and Sastry, K.S. 1973. Effect of oil sprays on the control of tomato leaf curl virus in the field. *Indian J. Agric. Sci.* **43**: 669-672.

Singh, J.P. and Varma, J.D. 1977. Effect of some insecticides and oils on incidence of yellow mosaic and grain yield of mungbean cv. Varsha. *Proc. Nat. Acad. Sci. India* **47**(B): 219-225.

Singh, H., G.S. Sandhu and G.S. Mari. 1971. Control of yellow mosaic virus in soybean, *Glycine max* (L.) Merrill by the use of granular insecticides. *Indian J. Entomol.,* **33**: 272-278.

Singh, J.P. 1980. Effect of virus diseases on growth components and yield of mungbean (*Vigna radiata*) and urdbean (*Vigna mungo*). *Indian Phytopath.,* **33**: 405-408.

Suteri, B.D. 1974. Occurrence of soybean yellow mosaic virus in Uttar Pradesh. *Curr. Sci.,* **43**: 689-690.

Thongmeerakom, P., K. Kittipokorn and P. Surin. 1981. Outbreak of mungbean yellow mosaic disease in Thailand. *Thailand J. Agric. Sci.,* **14**: 201-206.

Vandervekan, J. 1977. Oils and ohter inhibitors on non persistent virus transmission. In: K.F. Harris and K. Maramorosch (ed.) Aphids as virus vectors. Academic Press, pp. 435-454.

Vasudeva, R.S. 1958. Report of the division of Mycology and Plant Pathology. *Sci. Rep. Agric. Res. Inst.* New Delhi, 1955-1966, pp. 85-104.

Verma, H.N., P. Rastogi, V. Prasad, and A. Srivastava. 1985. Possible control of natural virus infection of *Vigna radiata* and *Vigna mungo* by Plant extracts. *Indian J. Pl. Path.,* **3**: 21-24.

Neem for the Management of Crop Diseases, 1995
Editor : V. Mariappan
Associated Publishing Co., New Delhi, India
pp. 145-155.

MANAGEMENT OF CHILLI MOSAIC VIRUS DISEASE BY USING PLANT PRODUCTS

Mariappan, V., Jayalakshmi, V. and Duleep Kumar Samuel, L.

Department of Plant Pathology, Tamil Nadu Agricultural University
Coimbatore-41003, Tamil Nadu

Chilli (*Capsicum annuum*) is an important spice crop grown in many countries. It is used extensively inculinary and medicinal purposes. The crop occupies an area of 892600 hectares with an annual production of 783300 tonnes. Tamil Nadu accounts for 29 per cent of the total production in the country (Paramaguru, 1992).

The crop is affected by more than 51 pathogens and among them 14 are virus diseases (Saha and Singh, 1988). Kandasamy (1972) reported that the chilli crop is effected by toabcco mosaic virus, potato virus, chilli mosaic virus, potato virus Y, tobacco etch virus, and alfalfa mosaic virus in Tamil Nadu. Among these potato virus Y has been reported to be prevalent either individually or in mixtures with other viruses (Dossa and Mungur, 1982). Kamara and Dubey (1975) reported the mosaic virus of chillies is transmitted by *Aphis craccivora* and*Myzus persicae*. Usually control of insect transmitted virus disease is achieved by controlling stylet-borne nonpersistent virus diseases by use of insecticides and by cultural practices but they have not yielded satisfactory results. Vandervekan (1977) discussed how oils inhibited aphid transmission of viruses, the efficiency of oil treatments and the possible field application of oils for the control of stylet-borne viruses. Bradley (1963) reported the control of potato virus Y by use of paraffin oil. Simons (1981) emphasized the use of oil formulations with naturally occurring antifeedants derived from plants for the control of insect transmitted virus diseases of crop plants.

Seed oils from plants like neem (*A. indica*) and custard apple (*Annona squamosa*) either individually or in mixtures have been reported to possess insecticidal properties against rice green leaf hopper (*Nephotettix virescens* Distant) and its transmission of rice tungro virus (Mariappan and Saxena, 1983, 1984). Chenulu (1984) obseved the prevention of transmission of a strain of water melon mosaic virus from infected plants to plants sprayed with various vegetable and mineral oils. The efficacy of plant extracts on the transmission of chilli mosaic viurs is reported in this paper.

Effect of plant extracts on the transmission of chilli mosaic virus when mixed with inoculum: The crude leaf extract preparations of these plants were mixed with the sap obtained from the chilli mosaic virus infected plants (at 9:1 v/v). Then the above preparations were inoculated by sap inoculation on the leaves of healthy chilli plants. Healthy chilli plants inoculated with the sap containing virus inoculum served as control. Among the 50 plant species tested many of them were effective in reducing the transmission of chilli mosaic virus. *Azadirachta indica* and *Bougainvillea spectabilis* were highly effective in reducing the chilli mosaic virus infection to an extent of 71.6 per cent over control. The reduction in the virus infectivity upto 68.3 per cent over control was achieved by the leaf extracts of *Ceiba pentandra, Cocos nucifera, Ipomoea cairica* and *Manilkara achras* (Table 1).

Table 1. The inhibition of chilli mosaic virus by leaf extracts posessing antiviral properties

Sl. No.	Name of the leaf extract	Inhibition of sap transmission 10 per cent
1.	*Adenathera pavonia*	66.6abc
2.	*Albizia lebbeck*	58.3defg
3.	*Annona squamosa*	66.6abc
4.	*Azadirachta indica*	71.6a
5.	*Bauthinia purpurea*	50.3defg
6.	*Borassus flabellifer*	28.3e
7.	*Bougainvillea spectabilis*	71.6a
8.	*Caesalpinia pulcherrima*	66.6abc
9.	*Carica papaya*	53.3fghi
10.	*Casuarina equisetifolia*	65.0abcd
11.	*Ceiba pentandra*	68.3ab
12.	*Citrus limon*	26.6o
13.	*Cocos nucifera*	68.3ab
14.	*Delonix regia*	41.6klm
15.	*Erythrina variegata* var *orientalis*	26.6o
16.	*Eucalyptus globulus*	56.6efg
17.	*Ficus benghalensis*	43.3kl
18.	*Ficus religiosa*	53.3fghi
19.	*Gliricidia sepium*	31.6mo
20.	*Hibiscus rosa-sinensis*	55.0efgh
21.	*Ipomoea cairica*	68.3ab
22.	*Ixora chinensis*	26.6o
23.	*Lagerstroemia speciosa*	65.0abcd
24.	*Leucaena leucocephala*	66.6abc
25.	*Madhuca indica*	16.0cdef
26.	*Manihot esculenta*	45.0jkl
27.	*Manilkara achras*	68.3ab
28.	*Millingtonia hortensis*	56.6efg
29.	*Morinda tinctoria*	60.0cdef
30.	*Moringa oleifera*	35.0mno
31.	*Morus alba*	26.6o
32.	*Nerium indicum*	53.3fghi
33.	*Parthenium hysterophorus*	51.6ghij
34.	*Peltophorum pterocarpum*	43.3kl
35.	*Phoenix pusilla*	46.6ijk
36.	*Pithecollobium dulce*	38.3lmn
37.	*Polyalthia longifolia*	56.6efg
38.	*Pongamia pinnata*	51.6ghij
39.	*Prosopis chilensis*	66.6abc

Sl. No.	Name of the leaf extract	Inhibition of sap transmission 10 per cent
40.	*Psidium guajava*	20.3o
41.	*Punica granatum*	45.0jkl
42.	*Ricinus communis*	18.3p
43.	*Samanea saman*	48.3hijk
44.	*Sesbania grandiflora*	25.0o
45.	*Syzygium cumini*	61.6bcde
46.	*Tabebuia rosea*	55.0edfgh
47.	*Tamarindus indica*	51.6ghij
48.	*Tectona grandis*	46.6ijk
49.	*Terminalia catappa*	45.0jkl
50.	*Ziziphus mauritiana*	59.3defg

All values in percentages. Mean of 3 replications. In a column means followed by same letter(s) are not significnatly different (P = 0.05) by DMRT. (Duleep Kumar Samuel, 1990)

Duleep kumar Samuel (1990) also observed that the flower extracts of 20 plant species when mixed with the infective sap obtained from chilli mosaic virus infected plants were inhibitory. A reduction in the virus infectivity upto 73.3, 71.6 and 68.3 per cent over control was recorded by the flower extracts of *Catharanthus roseus, Azadirachta indica* and *Bougainvillea spectabilis* (Table 2).

Table 2. The inhibition of chilli mosaic virus by flower extracts posessing antiviral properties

Sl. No.	Name of the flower extract	Inhibition of sap transmission over control 10 per cent
1.	*Albizia lebbeck*	58.3ef
2.	*Annona squamosa*	65.0cd
3.	*Azadirachta indica*	71.6ab
4.	*Bougainvillea spectabilis*	68.3bc
5.	*Caesalpinia pulcherrima*	60.0def
6.	*Calotropis gigantea*	61.6def
7.	*Carica papaya*	56.6f
8.	*Catharanthus reseusolia*	73.3a
9.	*Delonix regia*	51.6g
10.	*Eucalyptus globulus*	58.3ef
11.	*Hibiscus rosa-sinensis*	35.0h
12.	*Ixora chinensis*	38.3h
13.	*Lagerstroemia speciosa*	61.6def
14.	*Nerium indicum*	51.6g
15.	*Parthenium hysterophorus*	56.6f
16.	*Peltophorum pterocarpum*	63.3de
17.	*Punica granatum*	61.6def
18.	*Samanea saman*	58.3ef
19.	*Syzygium cumini*	63.3de
20.	*Ziziphus marutiana*	61.6def

All values in percentages. Mean of 3 replications. In a column means followed by same letter(s) are not significnatly different (P = 0.05) by DMRT. (Duleep Kumar Samuel)

Thus the leaf extracts and flower extracts when mixed with the virus inoculum were found to reduce the infectivity of chilli mosaic virus on sap inoculation. These observations have revealed the antiviral nature of the leaf extracts.

Efficacy of the plant extracts on the sap transmission of chilli mosaic virus when applied as pre-inoculation spray: The extracts of 18 plants species, flower extract of *Thespesia populnea* (10 per cent concentration) and the seed kernel extract of neem (*A. indica*) 5 per cent concentration) were sprayed on the leaf surface of healthy chilli plants. Neem seed kernel 5 kg was powdered and soaked in 100 l of water for eight hours. The filtrate was separated mixed with liquid soap at 1 ml/l and used for spraying. Three hours later the chilli mosaic virus inoculum was inoculated by sap inoculation method. Control plants were sprayed with water and inoculated with the virus inoculum only. A fortnight later the infected plants were counted and the percent reduction over control has been worked out for each treatment (table 3). Among these the neem seed kernel extract (NSKE) treatment recorded a maximum reduction in the chilli mosaic virus infection (66.6 per cent over control) followed by the leaf extracts of *Casuarina equisetifolia, Lantana camara, Bougainvillea spectabilis, Abutilon indicum, Clerodenderon inerme,* and *Cyprus rotundus* which recorded 60.0, 56.6, 53.3, 50.0 amd 50.0 per cent reduction over control respectively.

Table 3. Effect of pre-inoculation application of plant seed, flower and leaf extracts on chilli mosaic virus transmission by sap inoculation

Sl. No.	Name of the plant species	Percent disease reduction over control Mosaic*
1.	*Abutilon indicum*	50.0defghijl
2.	*Annona squamosa*	46.7efghijklmno
3.	*Aristolochia bracteata*	30.0abcde
4.	*Azadirachta indica*	66.6ijklmnopqrst
5.	*Bauhinia purpurea*	33.3abcdefg
6.	*Bougainvillea spectabilis*	53.3fghijklmnop
7.	*Casuarina equisetifolia*	60.0jiklmnopqrs
8.	*Clerodendron inerme*	50.0dfghijklm
9.	*Cyprus rotundus*	50.0defghijklmn
10.	*Eichhornia crassipes*	40.0abcdefgh
11.	*Lantana camara*	56.6fghijklmnopqr
12.	*Madhuca longifolia*	46.6defghij
13.	*Medicago sativa*	33.3abcdef
14.	*Morinda tinctoria*	23.3abc
15.	*Nerium indicum*	30.0abcd
16.	*Pongamia glabra*	43.0fghijklmnopq
17.	*Polyalthia longifolia*	40.0abcedefghi
18.	*Thespesia papulnea* (Flower)	20.0ab
19.	*Tridax procumbense*	20.0ab
20.	*Vinca rosea*	46.0defghijk

CD = 12.91 NSKE - Neem seed kernel extract.

* Mean of three replications. All values in percentages. In a column means followed by same letter(s) not significnatly different (P = 0.05) by DMRT. (Jayalakshmi, 1994)

Effect of plant extracts on the survival of aphid vectors: The leaf extracts of 17 plant species and the neem seed kernel extract were sprayed on healthy chilli plants. Three hours later viruliferous aphids (*Aphis craccivora* and *Myzus persicae*) were released at the rate of two insects per plant that had been treated with the preparations. For control treatment chilli plants treated with water spray were used. For each treatment 10 plants were used. The insects were confined to the plants by using myla cages. Three days later the dead insects were counted and the mortality rate were worked out (Table 4).

Table 4. Effect of plant products on survival of aphid vectors *A. craccivora* and *M. persicae*

Sl. No.	Plant Products	Percent insect mortality* over control	
		A. craccivora	*M. persicae*
1.	*Abutilon indicum*	56.6efghijklmn	26.7a
2.	*Acalypha indica*	33.3ab	53.3abcdefghi
3.	*Aristolochia bracteata*	50.0cdefghi	56.7bcdefghijklmn
4.	*Azadirachta indica* (NSKE)	70.0nopqur	70.0nopqr
5.	*Boerhaavia diffusa*	36.7abc	40.0abcde
6.	*Bougainvillea spectabilis*	50.0cdefghij	50.0abcdefghijkl
7.	*Casuarina equisetifolia*	53.3efghijklm	60.0ghijklmnop
8.	*Chenopodium amranticolor*	36.6abcd	43.3abcedfh
9.	*Clerodendron inerme*	56.7efghijklmno	50.0abcdefghijk
10.	*Cynodon dactylon*	46.7bcdefg	46.6bcdefgh
11.	*Cyprus rotundus*	56.7efghijklmnop	46.6bcdefghi
12.	*Datura metel*	56.7efghijklmnopq	50.0klmnopq
13.	*Lantana camara*	30.0a	40.0abcd
14.	*Mirabilis jalapa*	43.3abcdef	36.7abc
15.	*Parthenium hysterophorus*	50.0cdefghijk	53.3bcdefghijklmn
16.	*Polyalthia longifolia*	43.3efghijkl	50.0abcdefghijklm
17.	*Pongamia glabra*	46.7abcdefgh	40.0abcdef
18.	*Prosophis juliflora*	53.3abcde	36.6ab

* - Mean of three replications. NSKE - Neem seed kernel extract.
All the values in percentage. In a column means followed by same letter(s) are not significnatly different (P = 0.05) by DMRT.
(Jayalakshmi, 1994)

Among the treatments, neem (*A. indica*) seed kernel extract, followed by the leaf extracts of *Clerodenderon inerme, Cyprus rotundus, Datura metel, Casuarina equisetifolia* and *Prosphis juliflora* were relatively more effective than the other plant extracts and exerted high mortality (70.0 to 53.3 per cent) of the aphid species *A. cracivora*. In respect of *M. persicae* the treatments neem seed kernel extract, leaf extract of *Casuarina equisetifolia, Aristolochia bracteata* and *Acalypha indica* have been significantly more effective than the other treatments and recorded a mortality ranging from 70.0 to 53.3 per cent over control, The neem seed kernel extract has been highly effective in reducing the survival of both species of aphids Table 4 fig. 1.

Effect of plant products on the transmission of chilli mosaic virus by the aphid vectors: Healthy chilli plants of 40 days old were treated with leaf extracts

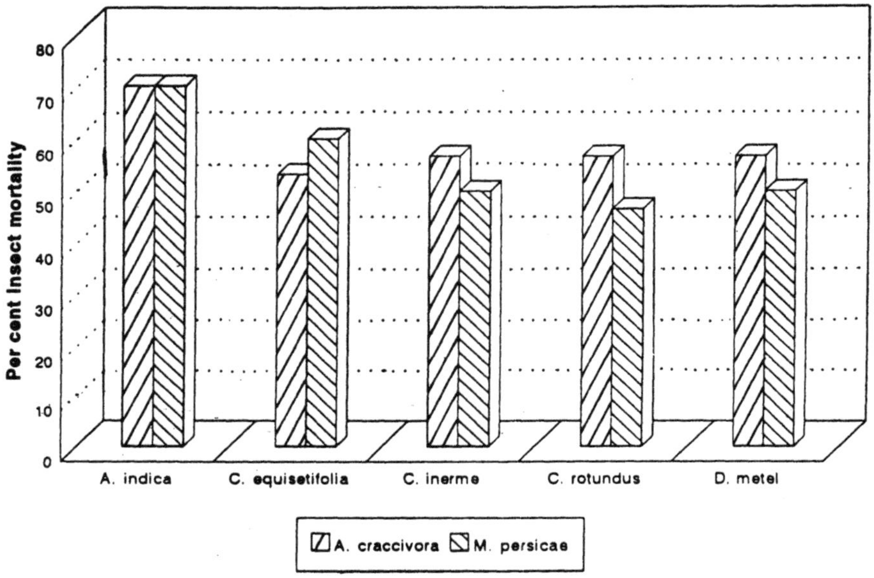

Fig. 1. Effect of plant extracts on survival of aphids (vectors), *A. craccivora* and *M. persicae*

of 17 plant species and neem seed kernel extract. For each preparation 20 seedlings were used. Three hours later the treated seedlings were exposed to viruliferous insects that had fed on chilli mosaic virus infected plants for 24 hours. The insects were allowed to feed on test plants (Chilli plants treated with plant extract preparations) at the rate of 2 insects per plant. Control plants were sprayed with water. The above experiment served as the pre inoculation treatment.

In another experiment the preparations (Plant extracts) were applied three hours after inoculation feeding by insects. The above experiment served as post-inoculation experiment.

A. craccivora: The transmission of chilli mosaic virus by *A. craccivora* was reduced significantly in respect of all treatments. In the preinoculation treatment there was a reduction of 86.6 per cent over control in the treatment of neem seed kernel extract. This was followed by *Prosophis juliflora* (80.0 per cent). In the post inoculation treatment maximum reduction (70.0 per cent) was observed in *P. juliflora* treatment (Table 5 and figure 2).

Table 5. Effect of plant products on transmission of chilli mosaic virus by *A. craccivora*

Sl. No.	Plant Products	Transmission per cent reduction over control	
		Pre inoculation*	Post inoculation*
1.	*Abutilon indicum*	73.3jklmno	53.0klmnop
2.	*Acalypha indica*	33.3a	16.0b
3.	*Aristolochia bracteata*	50.0bcdefg	23.3bc
4.	*Azadirachta indica* (NSKE)	86.6nopqur	33.3cde
5.	*Boerhaavia diffusa*	36.6ab	20.0ab
6.	*Bougainvillea spectabilis*	73.3jklmn	60.0mnopq
7.	*Casuarina equisetifolia*	66.6efghijk	46.6efghijklm
8.	*Chenopodium amaranticolor*	63.3defghijk	50.0jklmno
9.	*Clerodendron inerme*	46.6bcd	26.0bcd
10.	*Cynodon dactylon*	56.6cdefghi	36.6defgh
11.	*Cyprus rotundus*	43.0abc	46.6fghijklmn
12.	*Datura metel*	73.3jklmnop	40.0efghij
13.	*Lantana camara*	60.0bcdefghij	36.6defghij
14.	*Mirabilis jalapa*	53.3cdefgh	40.0efghijk
15.	*Parthenium hysterophorus*	66.6defghijklm	33.3cdefg
16.	*Polyalthia longifolia*	46.6bcde	43.3fghijkl
17.	*Pongamia glabra*	50.0bcdefg	33.3cdef
18.	*Prosophis juliflora*	80.0klmnopqr	70.0r

* - Mean of three replication. NSKE - Neem seed kernel extract.
All the values in percentage. In a column means followed by same letter(s) are not significnatly different (P = 0.05) by DMRT. (Jayalakshmi, 1994)

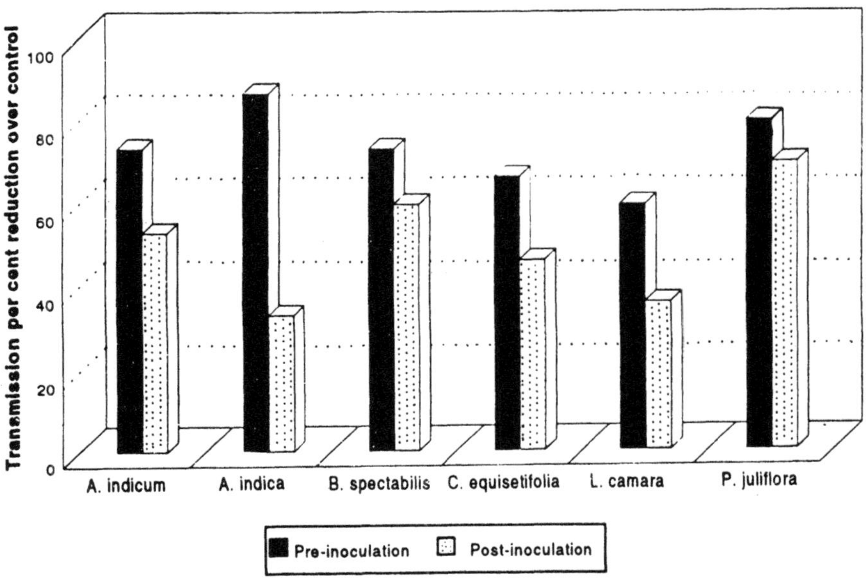

Fig. 2. Effect of plant extracts on transmission of chilli mosaic virus by *A. craccivora*

Table 6. Effect of plant products on transmission of chilli mosaic virus by M. persicae

Sl. No.	Plant Products	Transmission per cent reduction over control	
		Pre inoculation*	Post inoculation*
1.	*Abutilon indicum*	50.7cdefghijklmn	36.1jklmn
2.	*Acalypha indica*	46.6abcdefghijkm	23.3bcdefghi
3.	*Aristolochia bracteata*	36.6abcdef	30.0cdefghijklm
4.	*Azadirachta indica* (NSKE)	70.0defghijklmnopqr	53.3opqr
5.	*Boerhaavia diffusa*	33.3abcde	26.6bcdefghij
6.	*Bougainvillea spectabilis*	40.0abcdefghi	23.3bcedfgh
7.	*Casuarina equisetifolia*	46.6bcdefghi	40.0jklmno
8.	*Chenopodium amranticolor*	30.0ab	23.3bcdefg
9.	*Clerodendron inerme*	43.3abcdefg	30.0bcdef
10.	*Cynodon dactylon*	56.6cdefghijklmno	40.0jklmnop
11.	*Cyprus rotundus*	43.3bcdefghijk	23.3bcde
12.	*Datura metel*	63.3cdefghijklmnop	40.0jklm
13.	*Lantana camara*	46.6bcdefghijkl	30.0cdefghijkl
14.	*Mirabilis jalapa*	33.3abcd	20.0abc
15.	*Parthenium hysterophorus*	30.0abc	16.6bcd
16.	*Polyalthia longifolia*	40.0abcdegh	16.6a
17.	*Pongamia glabra*	26.6a	20.0ab
18.	*Prosophis juliflora*	66.6cdefghijklmnopq	50.0nopq

* - Mean of three replications. NSKE - Neem seed kernel extract.
All the values in percentage. In a column means followed by same letter(s) are not significnatly different (P = 0.05) by DMRT. (Jayalakshmi, 1994)

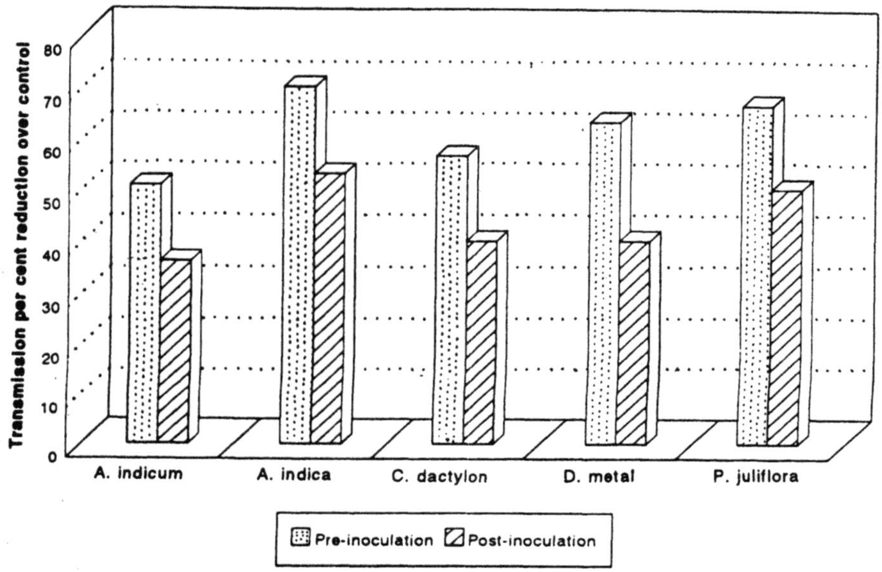

Fig. 3. Effect of plant extracts on transmission of chilli mosaic virus by *M. persicae*

M. persicae: The treatments NSKE (*A. indica* and *P. juliflora* recorded a maximum reduction of 70.0 and 66.0 per cent over control respectively in respect of pre inoculation treatment. In the post inoculation application also these two preparations were significantly more effective in reducing the transmission (Table 6 Fig. 3).

A field experiment was conducted with four selected plant extract preparations an insecticide and a virus suppressing chemical for the management of chilli mosaic virus disease during 1993. The treatments included neem (*A. indica*) seed kernel extract (5 per cent) leaf extracts of *Abutilon indicum, Bougainvillea spectabilis, Casuarina equisetifolia* (at 10 per cent), insecticide Monocrotophos (0.2 per cent) the chemical 'Salicylic acid' (0.2 per cent) and control (no treatment). The preparations were sprayed thrice on 7, 22 and 37 days at 500 l/ha. Chilli variety Co3 was used in the study. The percentage of virus infected plants and the yield recorded are furnished in table 7 (Fig. 4). From the results it may be seen percent disease incidence was low in all treatments (ranging from 17.0 to 23.3 per cent) as against the disease incidence recorded in control (30.2 per cent). Among the treatments neem seed kernel and salicylic acid recorded the least disease of 17.0 and 17.3 per cent respectively. With regard to the yield (Table 7) the treatment neem seed kernel extract yielded 11,076 kg/ha which was the maximum. This was followed by *B. spectablis*, which yielded 9877.2 kg/ha. The other treatments *A. indicum*, *C. equisetifolia*, monocrotophos and salicyclic acid) also recorded higher yields (ranging from 7599.0 to 7129 kg/ha) which were higher than the yield obtained from control plot 6775 kg/ha.

Table 7. Management of chilli mosaic virus disease under field condition

Sl.	Treatments	Percentage of disease	
		indicence*	Yield kg/ha*
T1	*Abutilon indicum*	22.7cde	7599.0abcde
T2	*Azadirachta indica* (NSKE)	17.0a	11076.4g
T3	*Bougainvillea spectabilis*	18.1abc	9877.2f
T4	*Casuarina equisetifolia*	21.6bcd	7161.1abc
T5	Monocrotophos	23.3def	7244.4abcd
T6	Salicylic acid	17.3ab	7125.0ab
T7	Control	30.2g	6775.8a

* Mean of three replications. In a column means followed by same letter(s) are not significnatly different (P = 0.05) by DMRT.

NSKE - Neem seed kernel extract.

Neem seed kernel extract has been found highly effective in reducing the chilli mosaic virus disease and in increasing yield of chilli.

Neem seed kernel extract has been found highly effective in reducing the transmission of chilli mosaic virus by sap inoculation as well as by the insect vectors (aphids). It was also found to be effective in reducing the insect survival. These results have clearly indicated that NSKE is both virucidal as well as insecticidal in

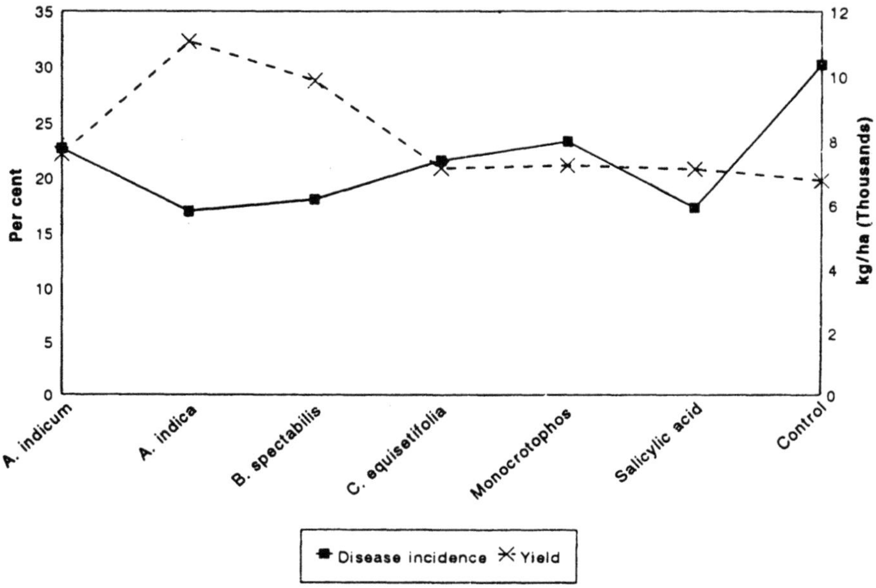

Fig. 4. Effect of plant extracts and chemicals on the incidence of chilli mosaic virus and yield under field conditions

property. The NSKE also has been found to be significantly effective fof field application and the farmers can use it relatively at a cheaper cost and increase their profit.

REFERENCES

Bradley, R.H.E. 1963. Some ways in which a paraffin oil impeded aphid transmission of PVY. *Canadian J. Microbiol.*, **9**: 369-380.

Chenulu, V.V. 1984. Plant virology in India-past, present and future. *Indian Phytopath.*, **37**(1): 1-20.

Dossa, M.I. and C. Mungur. 1982. The status of virus diseases of *Capsicum annuum* L in Mauritius. FAO. *Plant Prot. Bull.*, **30**: 151-156.

Duleep Kumar Samuel, L. 1990. Studies on chilli mosaic virus. M.Sc.(Ag.) Thesis, TNAU, Agricultural College and REsearch Institute, Madurai 147 p.

Kandasamy, T.K. 1972. Studies on virus disease of *Capsicum species in Tamil Nadu. Ph.D. Thesis, TNAU, Coimbatore.*

Mariappan, V. and R.C. Saxena. 1983. Effects of custard apple oil and neem oil on survival of Nephotettix virescens (Homoptera: Cicadellidae) and on rice tungro virus transmission. *J. Econ. Entomol.*, **78**: 573-576.

Mariappan, V. and R.C. Saxena. 1984. Effect of mixtures of custard apple oil and neem oil on survival of *Nephotettix virescens* and on RTV transmission. *J. Econ. Entomol.*, **77**(2): 519-521.

Paramaguru, P. 1992. Post harvest technology of chillies. *Spice India*, **1**: 16.

Saha, L.R. and H.B. Singh. 1988. Diseases of chilli and their management. *Int. J. Trop. Plant Dis.*, **6**: 135-143.

Simons, J.N. 1981. Innovative methods of control of insect transmitted plant viral diseases. In: *Vectors of disease agents, Interaction with plants, Animals and Man.* (eds.) J.J. McKelly (Jr.), Paregen Publishers, New York.

Vandervekan, J. 1977. Oils and other inhibitos of non-persistent virus transmission. In: *Aphids as virus vectors* (Eds.) K.F. Harris and K. Maramorosch. pp. 43-454. Academic Press, New York.

Neem for the Management of Crop Diseases, 1995
Editor : V. Mariappan
Associated Publishing Co., New Delhi, India
pp. 157-162.

STUDIES ON NEEM LEAF EXTRACT ON CUCUMBER MOSAIC VIRUS INFECTED TOBACCO AND CHILLI PLANTS IN TISSUE CULTURE

Rajyalakshmi Rao, D. and Rajam, M.V.

Department of Genetics, University of Delhi, South Campus, Benito Juarez Road
New Delhi-110021, India

A good deal of information is available on neem oil from *Azadirachta indica* and is known for its value as an effective antifungal (Jain and Agarwal, 1978), nematicidal (Wajid Khan *et al*, 1974), insecticidal (Srivastava *et al*, 1986) and antiviral agent i.e by checking infection through controlling vectors (Ketkar, 1976; Saxena and Justo, 1986, Mariappan *et. al*; 1985, Solsoly and Solsoly, 1987), except the work of Sangar and Dhingra (1982) on the effect of neem leaf extract on PVX. Present paper describes the effect of neem on the multiplication of cucumber mosaic virus on chilli and tobacco plants in various ways. Chilli saplings were raised aseptically from seeds on MS basal medium supplemented with neem leaf extract (50 ppm) before autoclaving, exercising leaf explants of chilli and tobacco was also carried out on the above MS modified medium to obtain calli respectively and also sraying it on tobacco plants *in viva* have been conducted.

MATERIALS AND METHODS

Preparation of materials and their inhibitory effect was tested as follows. The CMV maintained on tobacco (*Nicotiana tabacum* cv. White Burley) was used as inoculum by macerating young infected leaves in the ratio of 1:1 (w/v) of water.

Tender leaves of one year old neem plant were ground mixing one ml of distilled water for every gram of leaf tissue, extract was squeezed through two folds of muslin cloth and centrifuged at 3500 rpm for 30 min.

Different combinations of virus inoculum and the neem leaf extract in the proportion of 1:1, 1:5 and 1:10 were prepared to find out their effect on virus inhibition. Complete inhibition was observed at the ratio of 1:5 when assayed on *Chenopodium amaranticolor*.

RESULTS

Spraying the neem leaf extrat on upper surface of the leaves of the test plants (0, 24 and 48 hrs) before inoculation was effective in completely preventing virus infection for two months as symptoms regained after two months of treatment with variation in the severity. But virus infection was checked totally by three consecutive sprays with an interval of fifteen days. The dried and fresh neem leaves extract showed no difference in their action when sprayed on infected leaves of tobacco plants.

Tissue culture procedure: As Mn34 susceptible variety of chilli seeds were found to be more viable on the basis of germination percentage when raised aseptically on MS basal medium (Murashige and Skoog, 1962) and so these were used for all experiments.

Cucumber mosaic virus infected leaves of tobacco 15g was surface sterilised for 3 min. in .1% mercuric chloride and after giving three washes with sterile water for 5 min., leaves were crushed and ground in the sterilized pestle and mortar by adding equal amount of sterile distilled water (w/v) and sap was filtered through a double folds of muslin cloth. Chilli var Mn 34 seeds were surface sterilised as mentioned above and soaked in virus inoculum for 30 min. and then seeds were washed with distilled water to remove excess inoculum and cultured on MS basal medium (6 seeds per tube with 12 replicates) at a time. All cultures were maintained at 26 ± 1°C under a 16 hrs photoperiod of 2000 lux. All experiments were repeated thrice. Growth parameters were evaluated after 4 weeks of growth of cultures. Virus infectivity was assayed on *C. amaranticolor.* Inhibtion percentage was calculated by using the formula a-b/a × 100 (where a = the no. of lesions in control and b = no. of lesions in treated leaves). Results are depicted in Table 1.

Table 1. Effect of neem leaf extract (50 ppm) on the percentage germination size and the infectivity of chilli saplings when raised aseptically from CMV inoculated MN34 seeds

Treatments	% Germination* of Saplings		Size of Saplings (cms.)		Infectivity of CMV in Saplings	
	Range Variation	Mean	Range Variation	Mean	No. of Lesions	% inhibition
Healthy:						
Control	67-71	68	8-15.5	10.5	—	—
Treated	76-81	80	13-17.5	13.6	—	—
Diseased:						
Control	45-55	53	6-10	7.8	44	—
Treated	65-75	74	10-13.5	12.5	17	61

* Average of 12 replicates each with 6 seeds.

Leaf explants of virus infected chilli plants were excised from 4 weeks old aseptically grown neem treated saplings by cutting 6.6 mm leaf pieces in a laminar flow chamber and then placing them over MS callus and shoot regeneration medium respectively. Growth of the Calli was negative and also calli turned brown on the media except in control where callus was formed and it also retained infection.

Leaf explants of virus infected tobacco plants were excised from 6 weeks old plants grown in 6" pots with sterilised soil on MS medium callus and shoot regeneration medium supplemented with Neem leaves extract at 50 ppm before autoclaving to test their effect on CMV multiplication. The pH was adjusted to 5.8. Observations were taken after 30 days of growth and infectivity of the virus was assayed on *C. amaranticolor* using Latin Square Method and percentage inhibition was calculated. Data were subjected to analysis of variance (ANOVA) and results are presented in Table 2 and 3.

Table 2. Effect of neem leaf extract (50 ppm) on the growth of tobacco callus tissues infected with CMV on callus and shoot medium after 30 days of growth.

Growth	Callus		Shoot regeneration	
	Control	Treated	Control	Treated
Growth pattern	GG, LC	MG, CC	GG, LC	GG, CC
Wt. of callus (g)	1.278	0.987	1.302 (0.545)	1.228 (0.599)
No. of shoots	—	—	21	26

*GG - Good growth, MG - Moderate growth, LC - Loose callus, CC - Compact callus
Results are based on average of three experiments with 12 replicates each.

Table 3. Effect of neem leaf extract on the multiplication of CMV infected tobacco leaf explants when incorporated at 50 ppm in the MS callus and shoot regeneration medium.

Analysis of Variance variable INHPCT (inhibition %) by variable GRP (group)

Source	D.F.	Sum of Squares	Mean Squares	F Ratio	F Prob.
Between Groups	1	19507.0752	19507.0752	1286.0741	.0000
Within Groups	70	1061.7547	15.1679		
Total	71	20568.8299			

Group	Count	Mean	Standard Deviation	Standard Error	95 Pct Conf	Int for Mean
Callus	36	45.0125	4.8225	.8037	43.3808 To	46.6442
Shoot	36	77.9325	2.6607	.4435	77.0322 To	78.8328
Total	72	61.4725	17.0206	2.0059	57.4729 To	65.4722
Fixed	Effects	Model	3.8946	.4590	60.5571 To	62.3879
Random	Effects	Model		16.4600	−147.6715 To	270.6166
Random Effects Model-Estimate of Between Component Variance						541.4419

Group	Minimum	Maximum
Callus	35.7143	53.1250
Shoot	75.0000	83.3333
Total	35.7143	83.3333

Tests for Homogeneity of Variances
Cochrans C = Max. Variance/Sum (Variances) = .7666, P = .001 (appox.) Bartiett-Box F = 11.554, P = .001 Maximum Variance/Minimum Variance = 3.285

DISCUSSION

The results in Table 1 revealed that ability of neem to inhibit the multiplication of CMV in MS medium on chilli axenic plants was found to be 61 per cent. The mean

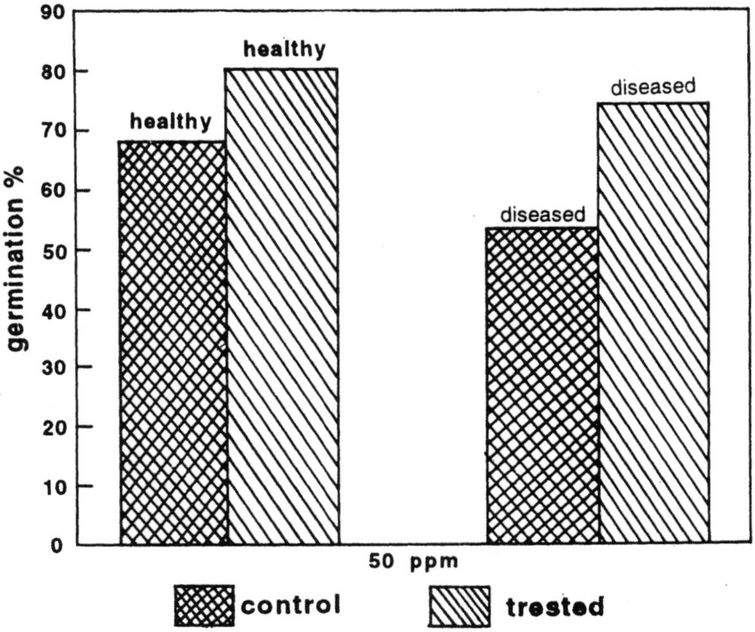

Fig. 1. Effect of Neem leaf extract on germination of Chilli seeds in MS medium

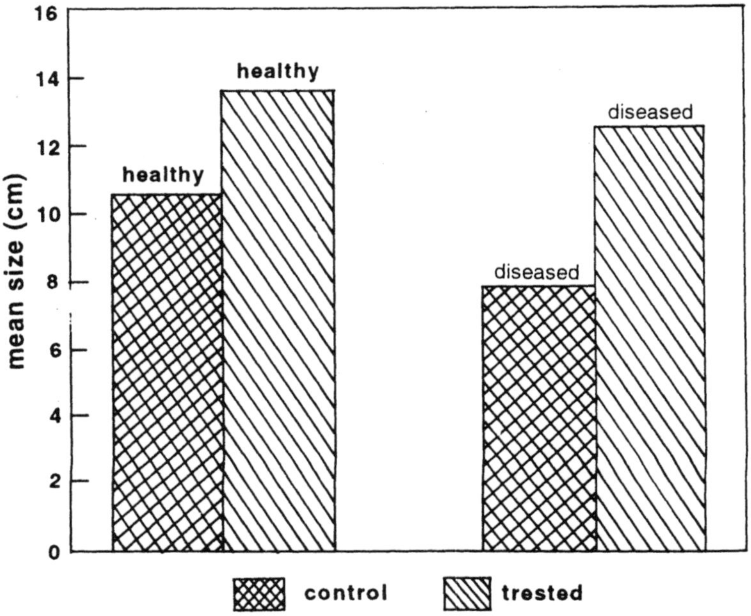

Fig. 1a. Effect of Neem leaf extract (50 ppm) on saplings of Chilli in MS medium

percentage germination of seeds with healthy uninoculated with virus and diseased inoculated with virus in treated was found to be 80 and 74 more as compared to the control healthy untreated and diseased untreated i.e 68 and 53 only. Whereas size of the sapling growth was also affected with neem and has been seen with treatment and without treatment in healthy and diseased mean size to be 13.6 and 12.5 and with control healthy and diseased 10.5 and 7.8 cms (Figure 1). This indicates that neem leaf extract is equally effective not only in reducing the viral infection also in enhancing the growth of the sapling. It also exhibited that it directly affects the growth of saplings irrespective of diseased or healthy, in fact more growth increase was observed with neem treated uninoculated than with inoculated saplings. But the effect of neem leaf extract on leaf explant of above virus infected chilli plants derived callus tissue grown on MS modified medium with 5mg/1 BAP when neem leaf extract was incorporated at 50 ppm before autoclaving failed either in enhancing the growth of the callus or multiplication of the virus in all the three times repeated experiment the calli turned brown. Reason could not be established. Further work is in progress to establish the cause of callus toxicity.

Data in Table 2 and 3 indicate that the effect of neem leaf extract (50 ppm) on the growth of tobacco callus tissue infected with CMV on callus and shoot regeneration medium after 30 days of growth, callus growth was good on shoot regeneration medium than on callus medium in both control and treated but there was not much difference in control and treated calli on callus medium. The percentage inhibition was more with treated calli on shoot regeneration medium than on callus medium i.e., 78% (S.EM– .4435) and 53% (S.EM– .8037) respectively. Calli was compact in all treated and was loose in untreated cases with good growth and did not turn brown. To test the effect of neem on CMV in non-differentiated and differentiated callus (NDC and DC) grown on shoot regeneration medium was considered. The MPI for 36 cases of NDC was 66.2 with SEM - 1.14, while that of DC was 75.8 with SEM - 0.6. The 'T' value for testing the significance of difference between two means was 7.46** (highly significant at 1% level) this infers that the inhibition was significantly higher in DC than in NDC in shoot induction medium. It was also found that neem not only checked the virus multiplication but also enhanced the shoot formation as compared to control by 25 percent. Precipitin test with CMV antiserum also revealed similar trend with NDC and DC treated infected calli (healthy NDC & DC treated/untreated calli –ve and diseased untreated and treated NDC/DC = ++/++– and ++–/+– respectively).

Spraying experiments with neem leaf extract showed complete inhibition with three consecutive sprays at the interval of 15 days at 1:5 ratio without any phytotoxic effect. Spraying neem leaf extract on test plants was also seen to inhibit PVX by Sangar and Dhingra (1982). Neem leaf extract at higher concentration was more effective in *in vivo* experiments and was not phytotoxic but in tissue culture it was not so, higher concentrations delayed the germination of seeds too, apart from turning brown the callus raised from leaf explants. Although both chilli and tobacco are solanaceous plants but the effect of neem extract differed in both cases former being more sensitive than the latter. It might be due to capsacin of chilli which has more role in interfering than nicotine of tobacco in growth of calli respectively.

ACKNOWLEDGEMENTS

Authors express their thanks to Dr. B.B. Singh, Head, Genetics Dept., NBPGR, New Delhi for providing seeds of MN 34 variety of Chilli and to University Grants Commission for awarding Research Associateship.

REFERENCES

Jain, P.C. and S.K. Agarwal. 1978. Note on the activity of some odouriferous organic compounds against keratophilic fungi. *Proc. Natl. Acad. Sci.*, **44**: 273-276.

Ketkar, C.M. 1976. Utilization of neem (*Azadirachta indica* A. Juss) and its by products. Report of modified Neem cake Manurial Project (1969-1976): 117-164.

Mariappan, V., R.C. Saxena, and K.C. Ling. 1985. Effect of custard-apple oil, neem oil and mixtures on survival of *Neophotettix virescens* and on rice tungro virus transmission. *Philipines Phytopathol.*, **19**: 36-43.

Murashige, T. and F. Skoog. 1962. A revised medium for rapid growth and bioassays with tobacco tissue culture. *Physiol. Plant.*, **15**: 473-497.

Sangar, R.B.S. and M.K. Dhingra. 1982. Potato virus x inhibitor from neem leaf extract. *J. Indian Potato Assoc.*, **9**: 148-149.

Saxena, R.C. and H.D. Justo, Jr. 1986. Effect of custard apple oil, neem oil and neem cake on green leafhopper (GLH) population on Tungro (RTV) infection (Philippines). *Int. Rice Res. Newsl.*, **11**: 25.

Solsoly, A.D. and T.S. Solsoly. 1987. Efficacy of neem seed oil in controlling bollworms, leafhoppers and aphids. Technical Report Cy 1985-86. Batac. Ilocos Norte (Phillipines). pp. 167-174.

Srivastava, K.M., N.S. Rana, Dwadash Shreni and B.P. Singh. 1986. Mechanism of inhibition of CMV by crude oil from Margosa *Aazadirachta indica* during inoculation with single apterous *Aphis gossypii. Indian Phytopath.*, **39**: 20-24.

Wajid Khan, M., M. Mashkoor Alam, A.M. Khan and S.K. Saxena. 1974. Effect of water soluble fractions of oil cake and bitter principles of Neem on some fungi and nematodes. *Acta Bot. Indica.*, **2**: 120-128.

Neem for the Management of Crop Diseases, 1995
Editor : V. Mariappan
Associated Publishing Co., New Delhi, India
pp. 163-173.

EFFECTIVENESS OF CHEMICAL COMPOUNDS FROM NEEM AND NEEM PRODUCTS TREATMENT IN CONTROLLING THREE IMPORTANT PLANT VIRAL DISEASES *IN VIVO* AND *IN VITRO*

Rajyalakshmi Rao, D. and Verma, V.S.

Division of Mycology and Plant Pathology, IARI, New Delhi-110012

Azadirachta indica A. Juss, a tree of paradise commonly known as 'neem' is planted all over India, because each and every part of the plant has been largely employed for medical purposes (Mitra, 1983), neem products are also useful as pest control agents for stored crops and standing crops (Jilani and Amir, 1987) and natural neem kills cockroaches and green house pests (Berberich, 1985). Apart from its uses the neem tree has natural resistance which is par excellence. It is an inhibitor of insect growth. Siddiqui (1942) reported the isolation of nimbin, nimbinin and nimbidin from neem oil. Extract from green leaves of neem has been found to inhibit potato virus X (Verma, 1970). In this paper studies carried out on the effectiveness of nimbin & nimbidin and two germicidal soaps 'Margo' and 'Neem' in controling PVX, TMV & CMV infection *in vitro and in vivo* have been discussed.

MATERIALS AND METHODS

In order to determine the effect of nimbin and nimbidin on a virulent strain of PVX (Verma 1970), the standard extract (SE) of this virus was prepared by grinding young infected leaves of *Nicotiana glutinosa* on which the virus was maintained, in a mortar. One ml of distilled water was used for each gram of leaves. The sap was obtained by filtering the pulp through a double fold of muslin cloth and was centrifuged at 3500 rpm for 30 minutes. Three concentrations of 250, 500, and 1000 ppm of nimbin and nimbidin by dissolving in a few drops of absolute alcohol were prepared. Different concentrations of both compounds were mixed with SE of the infective sap in equal proportions and allowed to stand for 15-30 min. Subsequently the mixture was used for inoculation of one half each of the leaves of *C.amaranticolor* Coste & Reyn; the other half was inoculated with virus inoculum mixed with equal volume of distilled water. The lesions were counted after 7 to 10 days and percentage inhibition was calculated according to the formula of Verma (1970). The results are presented in Table 1, 2, 3 and fig 1, 2.

Two brands of commonly available germicidal soaps viz. Margo (Made with Neem oil by Supercosmetics Pvt. Ltd, kanpur), Neem soap (contains neem oil and camphor, by Swastik surfactants Ltd, Bombay) were used. One percent stock solutions of both soaps were prepared and used for the experiments. TMV and CMV cultures were maintained in an insectproof glass house on *Nicotiana tabacum* cvs. Harrison Special and White Burley respectively. The standard extracts of these viruses TMV and CMV were prepared exactly same way as that of PVX mentioned above. In inoculum interference experiments, inocula containing virus and soap were prepared by mixing equal volumes of both and then diluting with distilled water to get .1, 1 and 10 ppm concentrations and mixing equal volumes of distilled water and virus then making dilutions with distilled water to get above concs, which served as control. Same experiment was repeated with both the viruses TMV and CMV with both the soaps respectively. Inocula were rubbed on carborundum-dusted leaves of local lesion hosts *Nicotiana glutinosa* and *Chenopodium amaranticolor* with a cotton bud by using half leaf method respectively. To compare TMV and CMV infection of soap-treated leaves to that of untreated leaves, a fully expanded leaves of *N. glutinosa* and *C. amaranticolor* were sprayed with a 1% solution of soap of margo and neem in distilled water and the opposite leaves with distilled water alone which served as control respectively. When leaves were dried after 20 minutes, inoculum was applied with a fresh cotton bud. Virus assay was calculated as mentioned in PVX. In another experiment, these soaps margo and neem reduced lesion number of TMV and CMV in *N. glutinosa* & in *C. amaraticolor* when these were pretreated with 1 per cent solution upto 24 & 48 hrs before virus inoculation. Same experiment after virus inoculation with soaps were not much effective in decreasing the virus infection of TMV & CMV. Data is presented in Table 4, 5, 6 and Fig 3 and 4.

In *vivo* experiments were conducted to find out the effect of these soaps on the multiplication of TMV & CMV callus tissues derived from tobacco leaf explants showing mosaic symptoms. Systemic hosts of TMV and CMV of *N. tabacum* cvs. Harrison Special and White Burley were established with 6" pots in the greenhouse respectively. Top few leaves were used in experiments. 1 per cent stock solution was prepared with each soap (Margo & Neem) and used in all experiments.

Callus cultures were aseptically raised in the Murashige and Skoog (1962) medium supplemented with 3% sucrose, 2mg/ml of IAA, 0.5 mg/ml kinetin, 4% 2-4-D and 15% coconut milk and 0.7% agar. The pH of the media was maintained to 5.8 to 6 prior to autoclaving at 15 lbs for 12 min. Before culturing, leaf explants were surface sterilized with 0.1% mercuric chloride for 2-3 min. and then washed with several changes of sterile water. They were cut into 0.6 X 0.6 cm pieces and placed on 10 ml medium in 25 X 150 mm pyrex tubes. All cultures were maintained at 24°C with a 16 hrs photoperiod. Test solutions of soaps Margo and Neem at 100 and 50 ppm were filtered through Whatman filter paper No.1 and incorporated in the MS medium. Distilled water incorporated in the medium served as control (Preliminary tests showed phytotoxic effect beyond this conc). The pH was adjusted to 5.8 before autoclaving. Later 30 days of growth, leaf derived TMV & CMV infected calli were cut into of almost equal size and transferred into these media.

After four weeks of growth, treated and untreated calli of TMV and CMV were freed from medium and their characteristic responses were observed (Table 7). Each sample was then macreated in sterile tissue grinder with an equal volume of distilled water (w/v) to get enough inoculum per sample and then the virus concentration of TMV and CMV was assayed on their respective local lesion hosts *Nicotiana glutinosa* and *Chenopodium amaranticolor* respectively (Table 8).

RESULTS

The results presented in Table 1 and fig. 1 depict that nimbin and nimbidin at 1000 ppm inhibited the virus by 44.1 and 62.8 per cent, respectively. The inhibition was markedly decreased at lower concentrations.

Table 1. Effect of Nimbin and Nimbidin on the Infectivity of PVX *in vitro*

Chemical	Per cent inhibition of the viral infectivity at different concentrations in ppm		
Compounds	250	500	1000
Nimbin	24.2	25.8	44.1
Nimbidin	40.6	44.2	62.8

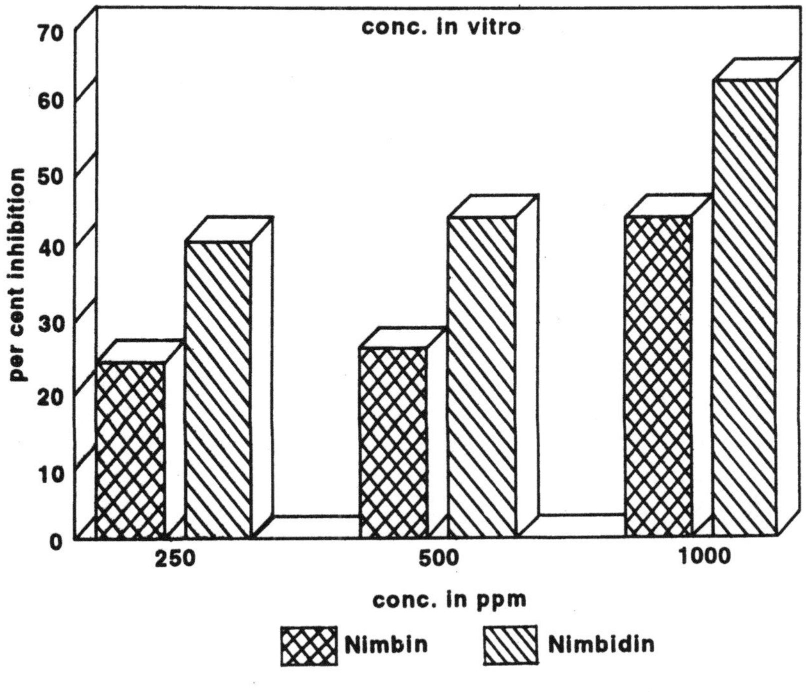

Fig. 1. Effect of Nimbin and Nimbidin on PVX

The results presented in Table 2, fig 2 show that highest inhibition (66.3 & 61.6 percent) of PVX infection was obtained when leaves were treated with nimbidin & nimbin 24 & 12 hrs before virus inoculation respectively, but at 6 hrs treatment both compounds decreased infectivity by 5.5 and 17.5 percent. In post-inoculation applications, highest inhibition of infection was obtained when the leaves were sprayed with both the compounds after 24 hrs of inoculation i.e., 42.5 and 35.3 percent respectively.

Table 2. Per cent inhibition of local lesion formation of young potato virus X on leaves of *Chenopodium amaranticolor* sprayed with nimbin and nimbidin at different times before and after inoculation

Chemical Compounds	Treatments in hours					
	Pre-inoculation			Post-inoculation		
	6	12	24	6	12	24
Nimbin	5.5	61.6	57.6	5.7	23.9	42.5
Nimbidin	17.5	55.7	66.3	22.7	28.3	35.3

The root treatments were applied to tomato seedlings. The roots of seedlings after thorough washing were dipped for 24 hrs in the above preparations and a

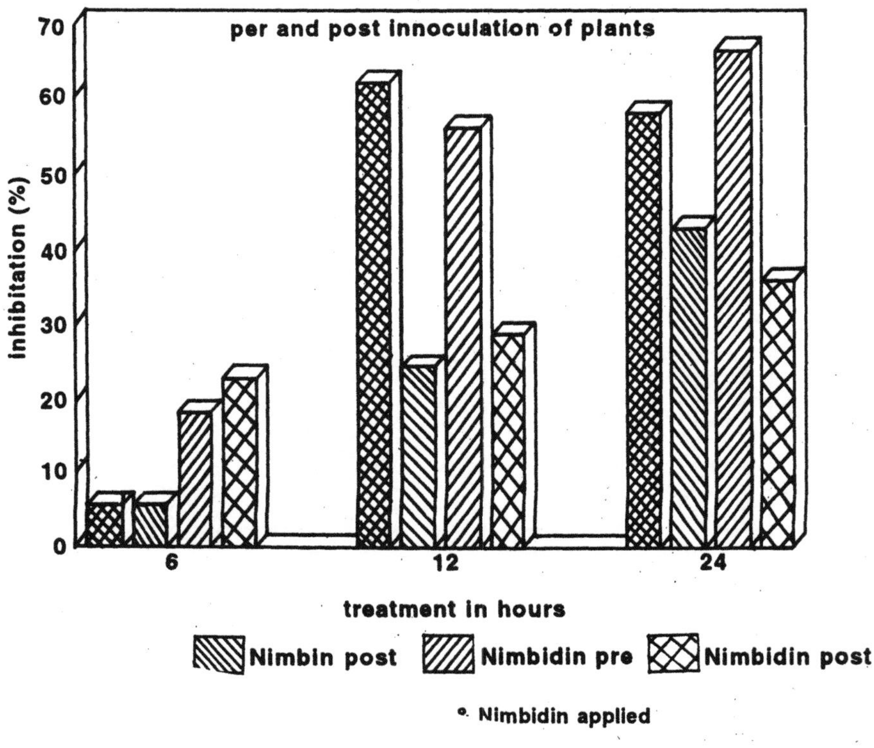

mixture of both chemicals at the same three concentrations was also used. The seedlings were then transplanted to pots. Another set of plants was kept for the same time in distilled water which served as control. After 24 hrs, all transplanted plants were inoculated with virus.

The Table 3, shows that there was no delay in the symptom expression by these chemicals. The symptoms in the treated plants were similar to that of control. About 20.0 and 10.8 per cent of virus inhibition *in vivo* was observed in 500 ppm nimbin & nimbidin respectively. The viral inhibition was significant at other concentrations. However the mixture of both chemicals did not increase the viral inhibition.

Table 3. Effect of root - dip treatments on viral infectivity *in vivo*

Chemical Compounds	Plant	Per cent inhibition *in vivo* at different concentrations in ppm 1000	500	150
Nimbin	Tomato	Phytotoxic	20.0	5.2
	Chenopodium	Phytotoxic	Phytotoxic	19.3
Nimbidin	Tomato	Phytotoxic	10.8	10.1
	Chenopodium	Phytotoxic	35.0	7.0
Mixture of	Tomato	Phytotxic	20.0	5.1
Nimbin and Nimbidin	Chenopodium	Phytotoxic	Phytotoxic	65.0

The above experiment was repeated with seedlings of *C. amaraticolor* and local lesion counts were made 10-12 days after inoculations (Table 3).

The results presented in Table 4 a and b show that Margo and Neem used at different concs. .1, 1 and 10 ppm inhibited the CMV more than the TMV. Maximum mean inhibition percentage (MIP) was shown with margo 86.7 and 74.4 and with neem 72.3 and 84.8 at 10 ppm on TMV and CMV respectively.

Table 4a. Effect of soaps solutions on the infectivity of TMV and CMV on *N. glutinosa* and *C. amaranticolor* plants.

Soap solns	Conc. in ppm	Mean % inhibition (SEM)	TMV	Mean % inhibition (SEM)	CMV
Margo	.1	53.43	(2.97)	54.83	(2.72)
Neem	.1	64.34	(4.03)	51.18	(3.03)
Margo	1	54.4	(3.39)	84.6	(1.75)
Neem	1	69.44	(2.05)	84.33	(3.03)
Margo	10	74.4	(1.28)	86.75	(3.57)
Neem	10	72.27	(1.12)	84.76	(3.04)

An analysis of 3-Way ANOVA of inhibition percentage by three factors - viruses, soaps and concentrations show that the F value for the explained variance is significant at 1 per cent level. Among the main effects the effect of soaps was not significant

Table 4b. Three way analysis of variance of the effect of soap solutions on the infectivity of TMV and CMV on *N. glutinosa* and *C. amaranticolor*

Source of Variance	Sum of Squares	DF	Mean square	F	Signif of F
Main Effects	19697.109	4	4924.277	659.257	.000
Virus	6516.795	1	6516.795	872.462	.000
Soap	.277	1	.277	.037	.847
Conc.	13180.037	2	6590.018	882.265	.000
2-way Interactions	10130.975	5	2026.195	271.265	.000
Virus Soap	151.126	1	151.126	20.233	.000
Virus Conc.	9761.135	2	4880.567	653.405	.000
Soap Conc.	218.714	2	109.357	14.641	.000
3-way Interactions	471.483	2	235.742	31.561	.000
Virus Soap Conc.	471.483	2	235.742	31.561	.000
Explained	30299.566	11	2754.506	368.771	.000
Residual	985.965	132	7.469		
Total	31285.531	143	218.780		

but main effects of virus and concentrations was statistically significant at 1 per cent level. The 2-Way Interaction of each of the three pairs-viruses/soaps, soaps/concentration and viruses/concentrations were found to be significant at 1 per cent level.

Table 5. Effect of Soaps on the Infectivity of TMV and CMV on *N. glutinosa* and *C. amaranticolor* leaves when dipped in 1 % soap solutions and allowed to dry for20 mins. before inoculation with virus.

Soap solns	Mean % inhibition (SEM) TMV		Mean % inhibition (SEM) CMV	
Margo	44.00	(2.97)	66.13	(2.72)
Neem	35.24	(3.05)	42.18	(2.07)

The 3-Way Interaction of all the three factors were also seen to be significant. The results in table 5 show that the highest inhibition of TMV and CMV was obtained (44 and 66 per cent) with margo and (35 and 42 per cent) with neem soaps when the leaves were dipped in these solutions (at 1 percent) for 20 mins. Whereas, when these were pretreated with 1 per cent solution upto 24 & 48 hrs before virus inoculation revealed marginal inhibition percentage difference with soaps as given in Table 6. Same experiment after virus inoculation with soaps were not much effective in decreasing the virus infection of TMV & CMV.

Table 6. Per cent inhibition of local lesion formation of viruses on *N. glutinosa C. amaranticolor* sprayed with 1% soap solutions at different times before and after inoculation

Chemical Compounds	Treatments in hours							
	Pre-inoculation				Post-inoculation			
	TMV		CMV		TMV		CMV	
	24	48	24	48	24	48	24	48
Margo	56	45	70	56	20	17	30	20
Neem	42	35	50	72	18	15	28	20

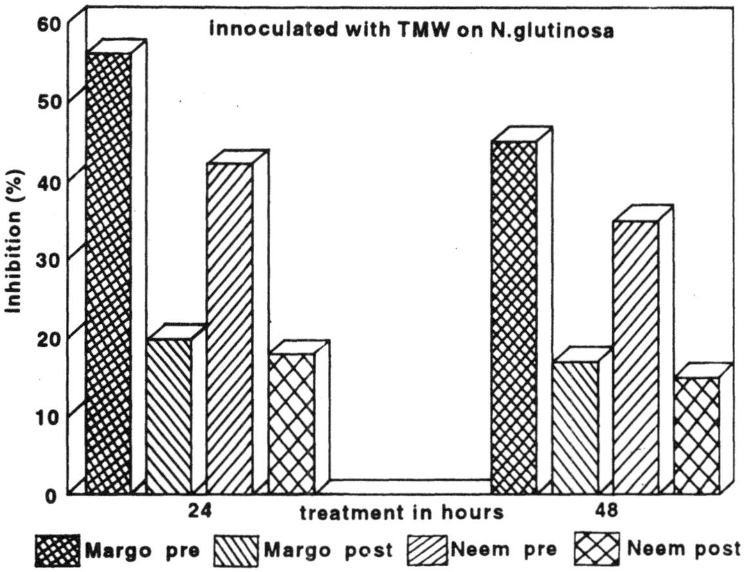

Fig 3. Effect of soap solns. when pre and post inoculated with TMV on *N. glutinosa*

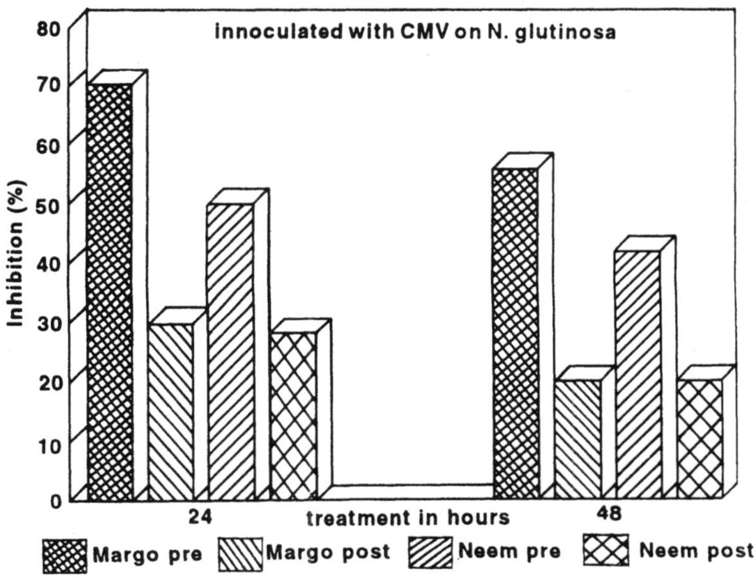

Fig 4. Effect of soap solns. when pre and post inoculated with CMV on *N. glutinosa*

When pieces of infected calli of TMV and CMV, induced from explants were transferred to fresh medium with different soap solutions of 50 and 100 ppm conc. respectively, three characteristic responses were observed (Table 7).

In both the cases change in colour, texture as well as growth in calli occurred. Maximum growth was noticed with Margo followed by neem and control. Trend remained the same with TMV and CMV except that the growth pattern was better in CMV than TMV infected calli.

Table 7. Growth pattern of TMV and CMV infected tobacco calli cultivated on media with soap solutions for 30 days.

Soap solns	Conc. in ppm	Growth pattern of infected calli					
			TMV			CMV	G
		C	T	G	C	T	G
Control	50	DB	LC	PG	DB	CC	PG
Margo	50	YB	CC	MG	BY	CC	MG
Neem	50	YB	LC	MG	BY	LC	MG
Control	100	DB	LC	MG	SB	LC	PG
Margo	100	YB	LC	MG	DB	LC	PG
Neem	100	B	LC	MG	SB	CC	PG

DB-Dark brown; YB-Yellowish brown; SB-Slight brown; BY-Brownish yellow; B-Brown; LC-Loose callus; CC-Compact callus; FG-Feeble growth; MG-Moderate growth; PG-Profuse growth; C-Colour; T-Texture; G-Growth.

The effect of soap solutions on the two viruses was different but not much i.e., mean inhibition percentage (MIP) in the case of Margo and Neem was approximately same. Concentrations of these solutions has marked impact on their inhibitory effect with TMV and CMV (Table 8)

Table 8. Effect of soap solutions on the multiplication of TMV and CMV in tissue culture.

Soap solns	Conc. in ppm	Mean % inhibition		Mean % inhibition	
		(SEM)	TMV	(SEM)	CMV
Margo	50	56.83	(2.66)	55.87	(3.64)
Neem	50	52.99	(2.18)	50.23	(3.68)
Margo	100	89.78	(2.66)	90.46	(1.80)
Neem	100	86.88	(2.47)	89.64	(2.08)

Three way analysis of variance of inhibitory effect achieved by different soap solutions at different concs. on the two viruses is presented in Table 9 also has evidenced effect of soaps like margo and neem derived from neem oil on viruses were statistically insignificant.

Table 9. Three way analysis of variance of the effect of soap solutions on the multiplication of TMV and CMV in tissue culture

Source of Variance	Sum of Squares	DF	Mean square	F	Signif of F
Main Effects	30532.160	3	10177.387	1563.246	.000
Virus	3.832	1	3.832	.589	.445
Soap	211.286	1	211.286	32.286	.000
Conc.	30317.042	1	30317.042	4656.396	.000
2-way Interactions	128.245	3	42.748	6.566	.000
Virus Soap	1.659	1	1.659	.255	.615
Virus Conc.	51.042	1	51.042	7.840	.006
Soap Conc.	75.544	1	75.544	11.604	.001
3-way Interactions	9.677	1	9.677	1.486	.226
Virus Soap Conc.	9.677	1	9.677	1.486	.226
Explained	30670.082	7	4381.440	672.989	.000
Residual	572.917	88	6.510		
Total	31242.999	95	328.874		

Over all MIP is 71.75 for viruses and average inhibition with CMV and TMV was found to be 71.5 & 71.9 respectively and MIP was almost same with the viruses. This is also confirmed by the 3-Way ANOVA, where main effects due to virus on the inhibition percentage was statistically insignificant. Effects of soaps and concentrations of soaps were significant at 1 per cent level. The interaction between virus and soap i.e., for both viruses (TMV & CMV) and soaps (Margo & Neem) appeared to be slightly better with Margo than with Neem but observed difference between the two soaps on the viruses is statistically insignificant. Increase in the concentration level of soaps increased the inhibition in the cases of both viruses and different in the effect of concentration on two viruses was highly significant at 1 per cent level.

DISCUSSION

The literature browsed indicates that the chemical compounds of neem were used against PVX for the first time. *In vitro* at a concentration of 1000 ppm, nimbidin and nimbin inhibited the virus by 62.8 and 44.1 per cent respectively. It is of interest that these chemicals were much more effective when applied before or 24 hrs after inoculation than when used at other intervals. In root treatments, the high percentage of inhibition may be due to higher accumulation of these chemicals in the leaves of *C. amaranticolor*, thereby inhibiting the viral infectivity and subsequent multiplication *in vivo*.

Inference can be drawn that the inhibition of the PVX particles by nimbin and nimbidin constituents of neem was only by their direct contact *in vitro*, while *in vivo* they did not inactivate the virus significantly.

From the data presented in Table 4-9, it is suggested that effect of germicidal soaps-Margo & Neem on the infectivity and multiplication of the viruses *in vitro* and *in vivo* reduced the TMV ana CMV virus infection. The results also reveal that reduction in TMV and CMV infection is primarily due to oil present in the soap and not soap as such. Germicidal soaps mainly Margo and Neem possess a pure oil of Neem, which has a property of insecticide in nature. Crude oil from *Azadirachta indica* is known for its value as an effective agent against some fungi (Jain & Aggarwal, 1978), nematodes (Wajid Khan *et al.,* 1974), Viruses (Verma 1970; Rai and Sethi, 1972; Reddy and Sethi, 1974; Mariappan *et al.*, 1985; Sangar and Dhingra, 1982; Saxena and Justo, 1986; Rajyalakshmi and Mahua Datta Gupta, 1993) and insects (Ketkar, 1978) etc. Neem soap also contains camphor, which appears to interfere with the action of neem individually as it is seen here that MIP with neem is found to be less than MIP with Margo on both the viruses (TMV and CMV).

According to Zinnen and Judith (1990) sole cause for the reduction of virus infection with soap solutions is pH. However, because soap, is composed of the salts of weak carboxylic acids, like Sodium dodecyl sulfate (SDS) alkaline (pH. 7.7 -9.9), whereas soaps used in the present studies were acidic (pH 5.8 -6). Since pH of the media of soap solutions were maintained constant. It is clear that only oil of the neem has a bearing on the reduction of virus multiplication. The mechanism of action may be a direct effect on the virus particle or the blockage of infection sites of the leaf with the oil of neem. Comparison of these two viruses also revealed that rod shaped virus (TMV) was more tolerant than spherical (CMV) to germicidal soaps.

REFERENCES

Berberich, S. 1985. Natural neem kills Cockroaches and green-house pests. Agriculture-Research-U.S.-Department of Agriculture, Agriculture-Research Service (USA) **34**: 13-14.

Jain, P.C. and S.K. Aggarwal. 1978. Note on the activity of some odoliferous organic compounds against keratophilic fungi. *Proc. Natl. Acad. Sci.,* **44**: 273-276.

Jilani, G. and P. Amir. 1987. Economics of Neem in reducing wheat storage losses-policy implications. *Science Technology* and *Development* (Pakistan), **6**: 1-9.

Ketkar, C.M. 1978. Utilization of neem (*Azadirachta indica* A. Juss) and its by products. Report of modified Neem cake Manurial Project (1969-1976): 117-164.

Mariappan, V., R.C. Saxena, and K.C. Ling. 1985. Effect of custard-apple oil, neem oil and mixtures on survival of *Neophotettix virescens* and on rice tungro virus transmission. *Philipines Phytopathol.,* **19**: 36-43.

Mitra, C.R. 1983. Neem. Indian Central Committee Hyderabad (Dn) India.

Murashige, T. and F. Skoog. 1962. A revised medium for rapid growth and bioassays with tobacco tissue culture. *Physiol. Plant.,* **15**: 473-497.

Rai, A. and M.S. Sethi. 1972. Screening of some plants for their activity against vaccines and fowl pox viruses. *Indian J. Anim. Sci.,* **42**: 1066-1070.

Rajyalakshmi Rao, D. and Mahua Datta Gupta. 1993. Effect of germicidal soaps on the multiplication of two viruses. Personal Communication.

Reddy, A.B. and M.S. Sethi. 1974. Antiviral effect of some independent plant extracts on vaccines and fowl pox viruses in chick embryo fibroblast cell culture. *Indian J. Expt. Biol.,* **12**: 572-574.

Sangar, R.B.S. and M.K. Dhingra. 1982. Potato virus x inhibitor from neem leaf extract. *J. Indian Potato Assoc.,* **9**: 148-149.

Saxena, R.C. and H.D. Justo, Jr. 1986. Effect of custard apple oil, neem oil and neem cake on green leafhopper (GLH) population on Tungro (RTV) infection (Philippines). *Int. Rice Res. Newsl.,* **11**: 25.

Siddiqui, S. 1942. A note on the isolation of three new bitter principle from the Nim oil. *Curr. Sci.,* **11**: 278.

Verma, V.S. 1970. Studies on a severe strain of Potato virus X from India. *Acta. Phytopath. Acad. Sci. Hung.*, **5**: 1970.

Verma, V.S. 1974. Chemical compounds from *Azadirachta indica* as inhibitors of potato virus X. *Acta. Microbiol. Poland*, B **6**: 9-13.

Wajid Khan, M., M. Mashkoor Alam, A.M. Khan and S.K. Saxena. 1974. Effect of water soluble fractions of oil cake and bitter principles of Neem on some fungi and nematodes. *Acta Bot. Indica*, **2**: 120-128.

Zinnen, T.M. and M.V. Judith. 1990. Insecticidal soap reduces infection by two mechanically transmitted plant viruses. *Plant Dis.*, **74**: 301-302.

C

Management of Diseases Caused by Nematodes

Neem for the Management of Crop Diseases, 1995
Editor : V. Mariappan
Associated Publishing Co., New Delhi, India
pp. 177-188.

NEEM IN MANAGEMENT OF PLANT NEMATODES

Saxena, S.K. and Saxena, Purnima

Botany Department, Aligarh Muslim University, Aligarh-202 002

Neem (*Azadirachta indica*) and its allied Bakain (*Melia azadirach*), belonging to the family Meliaceae, are widely grown in India from the time immemorial. Of these two important plants, the former has been referred widely in Vedic literature for its various health promoting and curative properties. Neem leaves have been reported to be used against various skin ailments (Rai, 1985) and infection of roundworms in man (Shah, 1982).

Plant-parasitic nematodes, attacking plants, also belong to the same phylum as the roundworms and cause considerable damage to crops. It is the antihelminthic (antinematode) properties of *neem* which prompted agricultural scientists to evaluate it against plant-parasitic nematodes.

All the plant-parasitic nematodes at least once in their life-cycles, pass through soil phase. In most of the cases the infective stage lives in soil. Therefore, almost all the treatments for control of nematodes were aimed towards this phase of nematode life-cycle. *Neem*/Bakain or their products have been added in soil as amendments. Various forms of these plants such as neem oil-cake, neem leaves, bark, shoot, root, fruit and gum and *Bakain* leaf etc. have been tested against nematodes as soil amendments or to their toxicity to development stage *in vitro*.

NEEM IN TOXICITY TO NEMATODES

(A) **Management of populations**

(i) *Neem oil-cake*: It has, by and large, been tested as soil amendments. To the ploughed, irrigated soil, with moisture equivalent ot water-holding capacity is added powdered neem oil cake. Crop is grown after allowing sometime for disintegration of oil-cakes.

Studies on the utility of neem oil-cake as organic amendments for controlling nematodes were started at Govindballabh Pant University of Agricultural Sciences and Technology, Pant Nagar (U.P) and Botany Department, Aligarh Muslim University, Aligarh (U.P) almost simultaneously (Singh and Sitaramaiah, 1966, Khan *et al*, 1966). In both the studies, it was demonstrated that neem oil cakes, besides other oil cakes, reduced the intensity of root-knot on vegetables. This treatment of soil with oil-cakes also has an adverse effect on other stylet-bearing nematodes. The neem oil-cake as soil amendments was aqually effective against rhizospheric pathogenic fungi (Khan *et al*, 1974). Soon after these reports, the results on efficacy of neem and other oil-cakes were confirmed against root-knot and other nematodes (Table 1).

Table 1. Efficacy of neem oil-cake reported against nematodes on various crops.

Nematode	Crop	Reference
Meloidogyne arenaria	Groundnut	Annon, 1993
M. incognita	Tomato	Goswami and Swarup (1971)
		Gowda (1972)
		Gowda and Sethy (1973, 1978)
		Mishra & Prasad (1978)
		Singh *et al.* (1980)
		Bhattacharya & Goswami (1987-1988)
		Alam (1989)
		Darekar *et al.* (1990)
	Eggplant	Alam (1989)
		Kirmani *et al.* (1978)
	Chilli	Alam (1989)
	Okra	Alam (1989)
		Reddy and Khan (1991)
	Blackgram	Ahmad *et al.* (1990)
	Chickpea	Pandey and Singh (1990)
	Green gram	Ravindra (1988)
	French bean	Srivastava and Singh (1990)
	Papaya	Routaray and Das (1988)
		Annon. (1993)
	Betelvine	Acharya and Padhi (1988)
	Tobacco	Desai *et al.* (1979)
	Davana *(Artemisia pallens)*	Haseeb and Butool (1991)
	Basil	Haseeb *et al.* (1988)
	Grapevine	Annon. (1993)
	Ginger	Annon. (1993)
M. javanica	Tomato	Singh and Sitaramaiah (1971)
		Srivastava *et al.* (1971)
	Eggplant	Srivastava *et al.* (1971)
	Okra	Singh and Sitaramaiah (1971)
		Zaiyd (1977)
Meloidogyne sp.	Tomato	Hameed (1970)
		Sen and Dasgupta (1989)
	Eggplant and Okra	Sen and Dasgupta (1989)
	Betelvine	Jagdale *et al.* (1985)
Rotylenchulus reniformis	Okra	Rao *et al.* (1987)
	Frenchbean	Padhi and Mishra (1987)
	Garden bean	Sundaram and Velayutham (1988)
	Cotton	Annon. (1993)
Tylenchorhynchus brassicae	Cabbage & Cauliflower	Khan (1969)
Hirschmanniella oryzae	Rice	Jonathan and Pandiarajan (1991)
		Mathur and Prasad (1973)

Nematode	Crop	Reference
Radopholus similis	Banana	Ravichandra and Krishnappa (1985)
		Annon. (1993)
Tylenchulus semipenetrans	Acid lime	Reddy *et al.* (1991)
Helicotylenchus dihystera	Garden gram	Vijaylakshmi and Prasad (1979)
General population of nematodes	Tomato	Alam *et al.* (1977)
		Ismail *et al.* (1976)
		Singh *et al.* (1980)
		Siddiqui and Alam (1991)
		Siddiqui and Saxena (1992)
	Potato	Akhtar and Alam (1991)
	Spinach	Alam and Khan (1974)
	Lentil	Gaur and Mishra (1989)
	Mung	Mishra and Gaur (1989)
	Wheat	Gaur and Prasad (1970)
		Mishra and Prasad (1974)
		Prasad *et al.* (1972)
		Vijayalakshmi and Prasad (1982)
	Triticale	Alam and Ashraf (1986)
	Grapevine	Darekar and Mhase (1990)
	Plum	Sharma (1991)
	Sugarcane	Jonathan *et al.* (1991)
	Berseem	Hasan and Jain (1984)

Water soluble fraction of neem oil-cakes as soil treatment was also found effective in managing nematode population (Sharma *et al.,* 1985). Disintegration of oil-cake in soil is a requirement for its efficacy. Singh *et al.* (1980, 1980a), while studying the irrigation requirement for proper disintegration of oil cake, suggested that neem oil cake as compared to other oil-cakes requires more irrigation water.

Both low and high temperatures did not favour decomposition of oil-cake. A temperature of 20°C was found optimum for effective decomposition of oil cake (Sitaramaiah, 1977). Therefore, the neem oil cake would be effective in reducing nematode population both in winter and summer, where temperature ranges from 10-35°C. The neem cake has been found equally effective in acid and alkaline soils (Alam, 1991).

(ii) *Neem leaves*: Neem leaves, as soil amendments not only act as source of green manuring but their disintegration products adversely affect the nematodes. Neem leaves have been tested as organic amendments against root knot nematodes (Table 2). This source has to be tested with other nematodes affecting the crops.

(iii) *Bakain*: Bakain has also yielded good results for management of nematodes but not as good as *neem.* In preliminary trials, oil-cakes, seed powder and leaf were incorporated in soil for control of root-knot nematode (*M. incognita*) infestation on tomato (Verma, 1986). Later, soil amendment with leaves of Bakain has given very effective control of root-knot nematode infestation on tomato and okra (Singh and Sitaramaiah, 1967) and of *M. incognita* on eggplant and tomato (Ahuja and Mukhopadhyay, 1987, Akhtar and Alam, 1990).

Table 2. Neem leaves as soil amendments against root-knot nematodes.

Nematode	Crop	Reference
Meloidogyne incognita	Tomato	Akhtar and Alam (1990)
		Rao and Bajaj (1984)
	Eggplant	Ahuja and Mukhopadhyay (1987)
		Pathak *et al.* (1988)
	Chilli	Akhtar and Alam (1990)
		Govindaiah *et al.* (1989)
	Mulberry	Govindaiah *et al.* (1989)
M. javanica	Tomato	Jain and Bhatti (1988)
		Singh and Sitaramaiah (1967)
		Zaki and Bhatti (1989)
	Okra	Singh and Sitaramaiah (1967)
		Zaiyd (1977)
	Chickpea	Gupta and Ram (1981)

All these treatments also reduced the general population of plant-parasitic nematodes on several crops (Haseeb *et al.*, 1978, and Akhtar *et al.*, 1990).

(iv) *Mechanism of action of Neem/Bakain Product as soil amendments*: Although at this stage it is very difficult to point out what mechanism(s) is/are operating in nematode management when neem oil cake or any other organic matter, is added as soil amendment. However, it has been suggested that there are four probabilities by which the organic amendment could help in combating nematode populations.

1. The nematoxic principles released by the disintegration of organic matter are directly toxic to nematodes.
2. The amendments result in changes in physical and chemical properties of soil favouring plant growth.
3. The organic matter dis-integration products when absorbed by plants induce resistance in plants against nematode attack.
4. The organic amendments cause increase in microflora of soil antagonistic to nematodes.

In order to elucidate the above, certain studies have been made in India which are reviewed below.

1. Disintegration products are directly toxic to nematode

Under laboratory conditions, neem and their products have been tested for their efficacy against nematodes. The water-soluble fractions caused mortality of nematodes and adversely influenced the hatching of larvae of nematodes. Neem oil-cake extract in water has also caused mortality of larvae of *M. incognita, Rotylenchulus reniformis, Tylenchorhynchus brassicae* and *Hoplolaimus indicus* (Khan *et al.* 1974b). Even extracts of decomposed oil cake were effective against nematodes. The longer the decomposition period, the more adverse effect on nematodes, does show that the nematoxic principles are released slowly during the course of decomposition.

Extracts of seed kernel, seed coat, seed extracts, leaf, shoot and root extracts were also effective against a variety of nematodes (Haseeb *et al.* 1978). The nematoxic effect of oil cake was retained even after removal of oil. The larvae of root-knot nematodes failed to penetrate the roots of tomato, resulting in poor galling (Mishra and Prasad, 1977).

Leaf and seed extracts of *Melia azadirach* have been reported toxic to nematodes (Lee, 1987, Siddiqui and Alam, 1985).

Neem cake during disintegration release a wide variety of compounds. The production of ammonia, formaldehyde and phenols during decomposition has already been documented (Alam *et al* 1978, 1979). These compounds are already known as nematoxic.

2. Changes in physical and chemical properties of soil

Not much work has been done in India in this direction but addition of organic matter adds to humus content of soil. This leads generally to more improvement of soil. The same principle operates when neem oil cake or any other organic product of neem such as leaf etc. is incorporated into the soil. Singh *et al.*, (1974) reported that when neem cake is added to soil, the absorption capacity of soil is increased. There is more availability of nutrients to plant, meaning thereby better growth and escape from nematode attack.

3. Disintegration products impart resistance to plants

Neem and its products have indicated systemic effect. The root knot development was reduced significantly when seedlings of plants were dipped in extracts of neem cake (Vijayalakshmi and Goswami, 1987, Abid and Maqbool, 1991); neem oil (Pradhan *et al.,* 1989) and neem leaf extract (Siddiqui and Alam, 1988). Seed treatment with an active principle of neem-azadirachtin protected the emerging seedlings (Siddiqui and Alam, 1988a). Foliar spray with neem leaf extract has also provided control of *Pratylenchus brachyurus* on maize (Egunjobi and Afolami, 1976).

These studies do indicate that neem extracts (oil cakes or of any part of plant) when tested as root-dip treatment, is absorbed by plants through roots, either provides tolerance directly to plants (phenols etc.) or triggers the metabolic activity leading to resistance in plants. The possibility of neem products, however, acting as elicitors of resistant mechanism cannot be ruled out.

Seed dressing with *neem* or their products has double role to play. The product, adhering to seed may have direct action on nematodes when seeds are sown in soil. At the same time with thin film of soil-water, the water soluble fraction is absorbed by plants imparting resistance to seedlings against nematode attack.

4. Increase in microflora antagonistic to nematode

By and large, when any organic matter is added in soil, it results in the development of wide variety of microoganisms including the ones which are antagonistic to nematode (Khan, 1969). The frequencey values and population of several saprophytic fungi including *Trichoderma viridae, Trichothecium roseum, Epicoccum purpurascens* and *Penicillium vermiculatum* increase in soil as a result of amendment with neem oil cake (Saxena, 1990). These fungi adversely affect the hatching of larvae of root-knot nematode. Similarly, the population of *Paecilomyces lilacinus* which is an established predator to namatode also increase due to soil amendment with neem oil cake (Khan and Husain, 1988; Reddy *et al.,* 1991).

B. Hatching and mortality of nematodes

Water soluble fractions/extracts of different derivatives of neem are toxic to nematodes. They are either nematicidal or nematostatic. These extracts also inhibited the hatching of larvae of root-knot nematode. Nimbin, thionemone, azadirachtim, nimbidic acid, kaepferal and quercetin (products from neem) have also been reported toxic to nematodes (Sitaramaiah *et al.*, 1974).

NEEM IN INTEGRATED PEST MANAGEMENT PROGRAMME ON NEMATODES (IPM)

Neem has been used, together with other nematode management tactics, in IPM to reduce the cost : return ratio and at the same time effectiveness. Initially neem cake has been tried with wheat straw (Prasad *et al.,* 1972), Sawdust (Singh *et al.,* 1983) and fertilizers (Gaur and Prasad 1970; Mishra and Prasad, 1978 and Prasad *et al.,* 1972). Neem oil cake, when mixed with other oil cakes, has also been very effective in reducing nematode numbers in soil (Alam *et al.,* 1982). This prompted industry to manufacture "Hind-o-Meal" - a mixture of *neem seed powder, bonemeal and NPK fertilizers* which was more effective than neem cake (Goswami *et al.,* 1988).

Deep ploughing of soil during May-June has been advocated as tillage practice for nematode control. This tillage practice together with soil amendment with neem oil cake has been very effective for nematode management (Saxena, 1990). Banana plantations treated with neem cake together with hot water remained free of infection of *Radopholus similis.*

Interculture with certain antagonistic plants has been suggested as a practice for containing nematode numbers. Mustard is a crop which acts against root-knot nematode. When potaotes are planted together with mustard (as interculture) in neem cake amended soil, the infestation of root-knot nematode was significantly low (Akhtar and Alam, 1991).

Systemic nematicides are applied on seeds as seed dressants. If such seeds are sown in nematode infested soil amended with neem cake the seedlings emerged from them are free of infection (Saxena, 1991). Treatment of soil with nematicides together with amendment of soil with oil-ckes has been tried in several studies in India. The treatment has been much more effective against nematodes and at the same time it reduced the cost : return ratio (Singh *et al.* 1980).

This is not only true with neem oil cake but with other neem products. Neem leaf mold together with sawdust as soil amendment was much more effective in reducing nematode numbers than neem leaf mold (Singh *et al.* 1982). The seedlings, grown in soil, manured with neem leaf mold remained free of nematode infestation (Jain and Bhatti, 1988). However, neem leaf mold was not as affective as neem oil-cake. But its integration with nematicides improved the effect of nematicides. Effective control of root-knot nematode was obtained on sugarcane when soil was treated with ethoprop and neem leaf mold. Root-dip treatment of tomato seedlings with phosphomidon before planting in soil amended with neem leaf mold was more effective against root-knot nematode than root-dip treatment with nematicides alone. Soil amendment with neem oil-cake together with soil solarization has been very efffective for nematode management (Siddiqui and Saxena, 1992).

NEEM IN OTHER METHODS OF NEMATODE CONTROL

Neem seedlings planted with field crops reduced nematode numbers in soil. Similar effect of interculture of seedlings of *Melia azadirach* was also observed. Root exudates of neem and *M. azadirach* were highely toxic to nematodes. The root exudates of neem were more toxic to nematodes than that of *M. azadirach* (Siddiqui and Saxena, 1987).

In view of the foregoing *neem* has a great potential for control of nematodes. Efforts are needed to work out the mechanism involved in such treatment. This would lead to pinpoint the exact active principle involved in killing of nematodes. Industries may come forward to give a dimension to formulate a nematicide of plant origin. However, more studies are needed in this direction.

REFERENCES

Abid M. and M.A. Maqbool. 1991. Effect of bare root-dip treatment in oil cakes and neem leaf extract on the root-knot development and growth of tomato and eggplant. *Pak. J. Nematol.*, **9**: 13-16.

Acharya, A. and N.N. Padhi. 1988. Effect of neem oil-cake and sawdust against root-knot nematode, *Meloidogyne incognita* on betelvine (*Piper betle* L.). *Indian J. Nematol.*, **18**: 105-106.

Ahmad, A., S.A. Tyagi and M.M. Alam. 1990. Efficacy of oil-seed cakes and nematicides against root-knot nematode attacking blackgram. *Sci. Khyber,* **3**: 165-170.

Ahuja, S. and M.C. Mukhopadhyay. 1987. Split application of nematicides to control root-knot nematode (*Meloidogyne incognita* and shoot and fruit borer (*Lencinodes orbonalis* Guen) in eggplant (*Solanum melongena*) seed crop. *Capsicum Newsl* **6**: 81-82.

Akhtar M. and M.M. Alam. 1990. Evaluation of nematicidal potential in some medicinal plants. *Int. Nematol. Network Newsl.*, **6**(1): 8-10.

Akhtar, M. and M.M. Alam. 1991. Integrated control of plant-parasitic nematodes on potato with organic amendments, nematicides and mix cropping with mustard. *Nematol. medit.*, **19**: 169-171.

Akhtar, M., S. Anwer and A. Yadava. 1990. Effects of organic amendments to soil as nematode suppressants. *Int. Nematal. Network Newsl.*, **7**(3): 21-22.

Alam, M.M. 1989. Control of root-knot and stunt nematodes with hornmeal, bone meal and oilseed cakes. Indian *J. Nematol.*, **19**: 166-170.

Alam, M.M. 1991. Control of plant-parasitic nematodes with oil seed cakes on some vegetables in field. *Pak J. Nematol.*, **9**: 21-30.

Alam, M.M. and S. Ashraf. 1986. Soil populations of plant parasitic nematodes infesting *Iriticale hexafloide* as influenced by organic soil amendments and nematicides. *Int. Nematol. Newsl.*, **3**(1): 19-20.

Alam, M.M. and A.M. Khan. 1974. Control of phytonematodes with oil-cake amendments in spinach field. *Indian J. Nematol.*, **4**: 239-240.

Alam, M.M., A.M. Khan and S.K. Saxena. 1977. Persistent action of oil-cakes and nematicides on the population of nematodes in field. *Botyn Kagaku*, **42**: 119-124.

Alam, M.M., A.M. Khan and Saxena. 1978. Mechanism of control of plant parasitic nematodes as a result of the application of organic amendments to soil. IV. Role of formaldehyde and acetone. *Indian J. Nematol.*, **8**: 172-174.

Alam, M.M., A.M. Khan and S.K. Saxena. 1979. Mechanism of control of plant parasitic nematodes as a result of the application of organic amendments to the soil. V. Role of phenolic compounds. *Indian J. Nematol.*, **9**: 136-142.

Alam, M.M., A.M. Khan, and S.K. Saxena. 1982. Relative toxicity of decomposed and undecomposed oil-cakes to plant-parasitic nematodes. *Acta Bot. Indica*, **10**: 124-127.

Annonymous 1993. Final biennial report on All India Coordinated Research Project for 1991-93.

Bhattacharya, D. and B.K. Goswami. 1987. Comparative efficacy of neem and groundnut oil-cakes with aldicarb against *Meloidogyne incognita* in tomato. *Rev. Nematol.*, **10**: 467-470.

Bhattacharya, D. and B.K. Goswami. 1988. Effect of different doses of neem and groundnut oil-cakes on plant growth characters and population of root-knot nematode. *Meloidogyne incognita in tomato. Indian J. Nematol.*, **18**: 125-127.

Darekar, K.S. and N.L. Mhase. 1990. Management of grapevine nematodes in field with neem cake and gramular nematicides. *Biored*, **1**: 101-103.

Darekar, K.S. N:L. Mhase and S.S. Shelke. 1990. Effects of placement of non-edible oil-cakes on the control of root-knot nematodes on tomato. *Int. Nematol. Network Newsl.*, **7**(1): 5-7.

Desai, M.V., H.M. Shah, S.N. Pillai and A.S. Patel. 1979. Oilcakes in control of root-knot nematodes. *Tobacco Res.*, **5**: 105-108.

Egunjobi, O.A. and S.O. Afolami. 1976. Effect of neem (*Azadirachta indica*) leaf extract on populations of *Pratylenchus brachyurus* and on the growth and productions of maize (*Zea mays*) in Nigeria. *Nematol. medit.*, **3**: 65-73.

Gaur, A.C. and S.K. Prasad. 1970. Effect of organic matter and inorganic fertilizer on soil and plant nematodes. *Indian J. Entomol.*, **32**: 186-188.

Gaur H.S. and S.D. Mishra. 1989. Integrated control of nematodes in lentil with aldicarb, neem cake and seed treatment with thimet and its residual effect on subsequent mung crop. *Indian J. Entomal.*, **51**: 283-287.

Goswami, B.K., M.M. Haque, S.D. Mishra and V. Majumdar. 1988. Comparative efficacy of Hind-o-meal, karanj, castor oil seed cakes and carbofuran against root-knot nematode in cowpea. *Ann. Agr. Res.*, **10**: 316-318.

Goswami, N. and G. Swarup. 1971. Effect of oil cake amended soil on the growth of tomato and root-knot nematode population. *Indian Phytopath.*, **24**: 491-494.

Govindaiah, N., Suryanarayana and D.D. Sharma. 1989. Effect of mulching of green leaves for the control of root-knot nematode in mulberry. *Indian J. Nematol.*, **19**: 25-28.

Gowda, D.N. 1972. Studies on comparative efficacy of oil cakes on control of root-knot nematode on tomato. *Mysore J. Agr. Sci.*, **6**: 524-525.

Gowda, D.N. and K.G.H. Setty. 1973. Studies on comparative efficacy of various organic amendments on control of root-knot nematode on tomato. *Mysore J. Agr. Sci.*, **7**: 419-423.

Gowda, D.N. and K.G.H. Setty. 1978. Comparative efficacy of neem cake and methomyl on the growth of tomato and root-knot development. *Curr. Sci.*, **7**: 118-120.

Gupta, D.C. and K. Ram. 1981. Studies on the control of *Meloidogyne javanica* infecting chickpea in different types of soil. *Indian J. Nematol.*, **11**: 77-80.

Hameed, S.F. 1970. Note on the effect of some organic additives on the incidence of root-knot nematode in tomato (*Lycopersicon esculentum* Mill). *Indian J. Agr. Sci.*, **40**: 207-210.

Hasan, N. and R.K. Jain. 1984. Effect of soil amendments on fodder production, photosynthetic pigments and nematodes associated with *berseem* (*Trifolium alexandrianum* L.) followed by bajra (*Pennisetum typhoides* L.). *Agr. Sci. Digest*, **4**: 12-14.

Haseeb, A., M.M. Alam, A.M. Khan and S.K. Saxena. 1978. Nematode population as influenced by soil amendments. *Geobios*, **5**: 152-155.

Haseeb, A. and F. Butool. 1991. Evaluation of different pesticides and oil-cake on the control of root-knot nematode, (*Meloidogyne incognita*) infesting on danana (*Artemisia pallens*). *Curr. Nematol.*, **2**: 201-204.

Haseeb, A.R. Pandey and A. Husain. 1988. A comparison of nematicides and oil-seed cakes for control of *Meloidogyne incognita on Ocimum basilicum. Nematropica*, **18**: 65-69.

Haseeb, A., B. Singh, A.M. Khan and S.K. Saxena. 1978. Evaluation of nematicidal property in certain alkaloid bearing plants. *Geobios*, **5**: 116-118.

Ismail, W., M.R. Kirmani, M.M. Alam and A.M. Khan. 1976. Control of phytophagous nematodes with oil-cake amendments around the roots of different varieties of tomato in field. *Proc. Nat. Acad. Sci. India*, **46**(B): 165-168.

Jagdale, G.B., A.B. Pawar and K.S. Darekar. 1985. Effect of organic amendments on root-knot nematodes infecting betel vine. *Int. Nematol. Network Newsl.*, **2**: 70.

Jain, R.K. and D.S. Bhatti. 1988. Integrated control of root-knot nematode, *Meloidogyne javanica* infecting tomato. *Indian J. Nematol.*, **18**: 30-34.

Jonathan, E.G., S.K. Krishnamorthi, M.I. Madharan and K. Muthukrishnan. 1991. Effect of organic amendments on the control of sugarcane nematode. *Bhartiye Sugar*, **16**(6):37: 39-40.

Jonathan, E.I. and P. Pandiarajan. 1991. Effect of organic amendments in controlling rice-root nematode. *Int. Rice Res. Newsl.*, **16**(2): 25.

Khan, A.M. 1969. Studies on plant-parasitic nematodes associated with vegetable crops in Uttar Pradesh. Final Technical Report, PL-480 scheme, 238 pp.

Khan, A.M., A. Adhami, Z.A. Siddiq and S.K. Saxena. 1966. Effect of different oil-cakes on hatching of larvae and on the development of root-knot caused by *Meloidogyne incognita* (Kofoid and White). Chitwood. *Inl. Symp. Plant Pathology*, New Delhi, p. 65. *On Plnt Disease Problems* pp. 562-568. (Eds. S.P. Raychaudhuri *et al.*). Indian Phytopathological Society, IARI, New Delhi, India.

Khan, A.M., M.M. Alam and R. Ahmad. 1974a. Mechanism of the control of plant-parasitic nematodes as a result of the application of oil cakes to the soil. *Indian J. Nematol.*, **4**: 93-96.

Khan, M.W., M.M. Alam, A.M. Khan and S.K. Saxena. 1974b. Effect of water soluble fractions of oil cakes and bitter principles of neem on some fungi and nematodes. *Acta Bot. Indica*, **2**: 120-128.

Khan, T.A. and S.I. Husain. 1988. Studies on the efficacy of seed treatment with pesticides, oil-cakes, neem leaf and culture filtrate of *Paecilomyces lilacinus* for the control of diseases caused by the presence of *Rotylenchulus reniformis, Meloidogyne incognita* and *Rhizoctonia solani* either individually or concomitantly on cowpea. *Indian J. Nematol.*, **18**: 192-198.

Khan, T.A., S. Nisar and S.I. Husain. 1991. Effect of water soluble extracts of certain oil-cakes on the development of root-knot disease of okra (*Abelmoschus esculentus*). *Curr. Nematol.*, **2**: 167-170.

Kirmani, M.R., S.K. Saxena and A.M. Khan. 1978. Growth and Development of root-knot on eggplant as influenced by some fungi. *Indian J. Nematol.*, **8**: 153-155.

Lee, M.J. 1987. The effect of extract of *Melia azadirach* on *Meloidogyne incognita. Quarterly J. Chinese Forestry*, **20**(4): 1-7.

Mathur, V.K. and S.K. Prasad. 1973. Control of *Hirschmanniella oryzae* associated with paddy. *Indian J. Nematol.*, **3**: 54-60.

Mishra, S.D. and H.S. Gaur. 1989. Control of nematodes infesting mung with nematicidal seed treatment and field application of aldicarb and neem cake. *Indian J. Ent.*, **51**: 422-426.

Mishra, S.D. and S.K. Prasad. 1974. Effect of soil amendments on nematodes and crop yields I. Oil seed cakes, organic matter, inorganic fertilizers and growth regulators on nematodes associated with which and their residual effect on mung. *Indian J. Nematol.*, **4**: 1-19.

Mishra, S.D. and S.K. Prasad. 1977. Effect of neem cake water extract on root-knot nematode infecting tomato. *Indian J. Ent.*, **39**: 228-231.

Mishra, S.D. and S.K. Prasad. 1978. Effect of soil amendments on nematodes and crop yields II. Oil seed cakes, organic matter and inorganic fertilizers at different levels of *Meloidogyne incognita. Indian J. Ent.*, **40**: 42-53.

Padhi, N.N. and R.P. Misra. 1987. Control of *Rotylenchulus reniformis* on Frenchbean (*Phaseolus vulgaris* L.). *Indian J. Nematol.*, 17.

Pandey, G. and K.P. Singh. 1990. Effect of organic amendments on soil microflora and nematode fauna with special reference to *Meloidogyne incognita* in chickpea. *New Agriculturist*, **1**: 65-70.

Pathak, A.K., B.S. Yadava, and J.S. Brar. 1988. Water hyacinth and neem leaves for the control of root-knot nematode, (*Meloidogyne incognita*) on brinjal. *Plant Dis. Res.*, **3**: 74-76.

Pradhan, P., S.N. Das and D.K. Dora. 1989. Effect of some organic oils on the infectivity of root-knot (*Meloidogyne incognita*) affecting tomato. *Indian J. Nematol.*, **18**: 117-118.

Prasad, S.K., S.D. Mishra and A.C. Gour. 1972. Effect of soil amendments on nematodes associated with wheat followed by mung and maize. *Indian J. Ent.*, **34**: 307-311.

Rai, M.K. 1985. Plants used as medicines by tribals of chhindwara district (M.P.). *J. Econ. Tax. Bot.*, **7**: 385-387.

Ram, K. and D.C. Gupta. 1980. A note on the efficacy of fresh neem leaf extract in the control of *Meloidogyne javanica* infecting chickpea (*Cicer arietinum*). *Indian J. Nematol.*, **10**: 96-98.

Ram, K. and D.C. Gupta. 1981. Studies on the control of *Meloidogyne javanica* infected chickpea (*Cicer arietinum*). *Haryana Agr. Univ. J. Res.*, **11**: 77-81.

Rao, A.B. V.K., N.N. Padhi and A. Acharya. 1987. Effect of different nematicides, oil cakes and urea in the control of *Rotylenchus reniformis* on okra. *Indian J. Nematol.*, **17**: 171-173.

Rao, B.H.K. and R. Bajaj. 1984. Effect of decomposing organic matter on plant growth and the incidence of root-knot nematode in *Lycopersicon esculentum. Indian Phytopath.*, **37**: 160-164.

Ravichandra, N.G. and K. Krishnappa. 1985. Effect of various treatments, both individually and in integration in controlling burrowing nematode, *Radopholus similis* infesting banana. *Indian J. Nematol.*, **15**: 62-65.

Ravindra, H. 1988. Studies on root-knot nematode, *Meloidogyne incognita* on green gram (*Vigna radiata* L.). Wilzeck (Thesis Abstract). *Mysore J. Agr. Sci.*, **22**: (1 Suppl.) 201-202.

Reddy, P.P. and R.M. Khan. 1991. Integrated management of root-knot nematodes infecting okra. *Curr. Nematol.*, **2**: 115-116.

Reddy, P.P., R.M. Khan and M.S. Rao. 1991. Integrated management of Citrus nematode, *Tylenchulus semipenetrans* using oil-cakes and *Paeciliomyces lilacinus. Afro-Asian J. Nematol.*, **1**: 221-222.

Routaray, S.B.N. and S.N. Das, 1988. Population management of root-knot nematode, *Meloidogyne incognita* on papaya. *Int. Nematol. Newsl.*, **5**(3): 25-27.

Saxena, S.K. 1990. Annual Report of All India Coordinated Research Project on nematodes, A.M.U., Aligarh.

Saxena, S.K. 1991. Annual Report of All India Coordinated Research Protection nematode, A.M.U., Aligarh.

Saxena, S.K. 1993. Role of soil solarization in Integrated nematode management. *Proc. Indian Science Congress Assoc.* Abst.

Sen, K. and M.K. Dasgupta. 1989. Control of root-knot nematodes (*Meloidogyne* spp.) with organic amendments and soil nematicides on some successive crops. *Horticultural J.,* **2**: 48-54.

Shah, N.L. 1982. Herbal folklore medicine in North India. *J. Etemopharmacol.,* **6**: 293-301.

Sharma, G.C. 1991. Yield losses in plum (*Prunus domestica*) due to nematodes. *Curr. Nematol.,* **2**: 171-174.

Sharma, H.L., O.P. Vimal and D. Prasad. 1985. Nematicidal and manurial value of some oil-cakes. *Int. J. Tropical Pl. Dis.,* **3**: 51-55.

Siddiqui, M.A. and M.M. Alam. 1985. Evaluation of nematicidal properties of different parts of margosa and Persian lilac, *Neem Newsl.,* **2**: 1-4.

Siddiqui, M.A. and M.M. Alam. 1988. Control of root-knot and reniform nematodes by bare root-dip in leaf extracts of margosa and Persian lilac. *Zeitschrift fur Pflanzenkrankheiten und Pflanzenschuldz,* **95**: 138-142.

Siddiqui, M.A. and M.M. Alam, 1988a. Effect of seed treatment with azadirachtin on root-knot development on, and growth of, some vegetables. *Tests* Agrochem Cult. **9**: *Ann. appl. Biol.,* **112** (Supp): 20-21.

Siddiqui, M.A. and M.M. Alam. 1991. Efficacy of oil seed cakes and nematicides against plant-parasitic nematodes attacking tomato in relation to ploughing. *J. Bangladesh Acad. Sci.,* **15**: 23-29.

Siddiqui, M.A. and S.K. Saxena. 1987. Effect of interculture of margosa and Persian lilac with tomato and eggplant on root-knot and reniform nematodes. *Int. Nematol. Network Newsl.,* **4**: 27-29.

Siddiqui, M.A. and S.K. Saxena. 1992. Effect of soil solarization together with oil-cakes/nematicides on nematode population and plant growth of tomato. *Proc. Nat. Acad. Sci. India.,* **62** B: 124-127.

Singh, R.P., A.M. Khan and S.K. Saxena. 1974. Effect of oil-cakes on the movement of nematicides in soil. Nematol. medit., **7**: 209-215.

Singh, R.S. and K. Sitaramaiah. 1966. Incidence of root-knot of okra and tomato in oil-cake amended soil. *Plant Dis. Reptr.,* **50**: 668-672.

Singh, R.S. and K. Sitaramaiah. 1967. Effect of decomposing green leaves, sawdust and urea on the incidence of root-knot of okra and tomato. *Indian Phytopath.,* **20**: 349-355.

Singh, R.S. and K. Sitaramaiah. 1971. Control of root-knot through organic and inorganic amendments of soil. Effect of oil-cakes and sawdust. *Indian J. Mycol. Pl. Pathol.,* **1**: 20-29.

Singh, S.P., M. Ahmad, A.M. Khan and S.K. Saxena. 1980. Effect of seed treatments with certain oil-cakes or nematicides on the growth of tomato and rhizosphere population of nematodes and fungi. *Nematol. medit.,* **8**: 193-198.

Singh, S.P., A.M. Khan and S.K. Saxena. 1980a. Effect of watering and mode of application of oil cakes and nematicides on their efficacy in controlling root-knot nematode on tomato. *Acta. Bot. Indica,* **88**: 193-195.

Singh, S.P., A.M. Khan and S.K. Saxena. 1980b. Effect of nematicides and oil cakes separately and in mixtures on plant-parasitic nematodes. *Geobios,* **7**: 11-113.

Singh, S.P., A.M. Khan and S.K. Saxena. 1980c. Effect of watering and mode of application of oil cakes and nematicides in controlling plant nematodes. *Geobios,* **7**: 145-148.

Singh, S.P., V. Pant, A.M. Khan and S.K. Saxena. 1982. Effect of sawdust with different nitrogen sources on the growth of tomato and on rhizosphere population of nematodes and fungi. *Indian Phytopath.,* **36**: 417-421.

Singh, S.P. and V. Singh. 1988. Evaluation of oil-cakes and nematicides for the control of *Meloidogyne incognita infecting eggplants. Indian J. Nematol.*

Sitaramaiah, K. 1977. Effect of temperature on decomposition of margosa cake and on survival of *Tylenchorhynchus elegans. Indian J. Nematol.,* **7**: 167-169.

Sitaramaiah, K., R.S. Singh and B.P. Pal. 1974. Effect of extracts and distillates of amended soil on activity of *Meloidogyne javanica. Indian J. Mycol. Pl. Pathol.,* **4**: 138-144.

Srivastava, A.K. and R.B. Singh. 1990. Effect of organic amendment on interaction of *Macrophomina phaseolina* and *Meloidogyne incognita* on Frenchbean (*Phaseolus vilgaris*). *New Agriculturist*, **1**: 99-100.

Srivastava, A.S., R.C. Pandey and S. Ra. 1971. Application of organic amendments on the control of root-knot nematode, *Meloidogyne javanica (Trcub)*. *Labdev J. Sci. Tech.*, **9B**: 203-205.

Sundaram, R. and B. Velayutham. 1988. Relative efficacy of some insecticides and neem cake in the control of *Rotylenchulus reniformis* and *Helicotylenchus dehystera* affecting garden bean.

Verma, R.R. 1986. Efficacy of organic amendments against *Meloidogyne incognita* infesting tomato. *Indian J. Nematol.*, **16**: 105-106.

Vijayalakshmi, K. and B.K. Goswami. 1987. Effect of root dip treatment of tomato seedlings of aqueous extracts of some oil seed cakes on root-knot nematode infestation. *Ann. Agr. Res.*, **8**: 168-171.

Vijayalakshmi, K. and S.K. Prasad. 1982. Effect of some nematicides, oil seed cakes and inorganic fertilizers on nematodes and crop growth. *Ann. Agr. Res.*, **3**: 133-139.

Zaiyd, M. 1977. Effect of organic soil amendments on the incidence of root-knot nematode (*Meloidogyne javanica*) on bhindi plants. *Proc. Bihar Acad. Agr. Sci.*, **25**: 23-26.

Zaki, F.A. and D.S. Bhatti. 1989. Phytotherapeutic effect on some plant leaves on *Meloidogyne javanica* infecting tomato plants. *Nematol. medit.*, **17**: 71-72.

Neem for the Management of Crop Diseases, 1995
Editor : V. Mariappan
Associated Publishing Co., New Delhi, India
pp. 189-199.

EFFICACY OF NEEM AND NEEM BASED PRODUCTS ON MANAGEMENT OF PARASITIC NEMATODES OF CROP PLANTS

Sivagami Vadivelu

Professor & Head, Department of Nematology, CPPS, TNAU
Coimbatore-641 003

INTRODUCTION

The medicinal properties on neem *(Azadirachla indica)* including its anthelminthic properties are well established and doumented. The deoiled cakes of neem has good manural value in al the tropical and subtropical countries. In addition to organic matter, neem cake contains nitrogen (55 to 7.1 per cent), phosphorous (1.1 per cent) and potassium (1.5 per cent).

Different parts of the tree has been recognised to possess insecticidal, nematicidal and fungicidal activity, since, the middle of 20th century. Neem cake as such has been reported to have selective toxic effect on plant parasitic nematodes. It also promotes nematode antagonistic microflora in soil ecosystem.

Based on the nematicidal efficacy of neem, experiments were conducted at Tamil Nadu Agricultural University on different crops like banana, grapevine, tomato, onion, betelvine etc.

Banana: In banana the deoiled cakes of neem, castor groundnut, pungam and along with compost were tried @ 500 kg/sucker in the variety Robusta. Carbofuran 3G @ 40g/sucker was also tried. The population of nematode at harvest both in soil and root and bunch weight of banana were recorded. The observations appended in Table 1. indicates that the neem cake applied suckers registered 172 nematodes per 250g. soil and 82 per 5g. roots, which is on par with pungam and castor cakes.

In the variety Rasthali urea (110/plant) and neem coated urea 110g/plant and 55g/plant were tried. The nematode population was assessed both at flowering and at harvest. The population data appended in Table 2 indicate that the reduction in the population was (85.2 and 78.0 percent in soil and root) with neem cake coated urea @ 100g/plant when compared to half the dose of neem cake coated urea with 67.9 and 52.7 per cent reduction respectively over normal nitrogen supplied through urea alone.

Table 1. Banana (Robusta) - Organic amendments on nematode and plant - 1985-1986 (3 replications)

Sl. No.	Treatments	Nematode population at harvest		Duration (days)	Bunch weight (Kg)
		Soil (250g)	Root (5g)		
1.	Neem cake 500g/sucker	172[a]	82[a]	378[d]	15.9[ab]
2.	Compost "	245[c]	149[b]	368[b]	15.5[ab]
3.	Castor cake "	166[ab]	92[ab]	384[e]	14.5[a]
4.	Groundnut cake "	136[a]	71[a]	354[a]	19.4[c]
•5.	Pungam cake "	176[b]	75[a]	371[bc]	15.0[a]
6.	Furadon 40g sucker	121[a]	57[a]	367[b]	19.0[c]
7.	Control	892[d]	315[c]	363[b]	12.3[a]
	C.D. (P = 0.05)	39.15	29.61	5.86	3.08

Subramaniyan S. and Selvaraj (1986)

Table 2. Banana (Rasthali) - Neem-cake coated urea on nematodes - 1985-1986 (3 replications)

Sl. No.	Treatments	Nematode population			
		At flowering		At harvest	
		Soil (250g)	Root (5g)	Soil (250g)	Root (5g)
1.	Neem cake coated urea at 110g/plant	17.3	11.0	63.0	35.6
2.	Neem cake coated urea at 55g/plant	37.5	24.0	87.6	65.8
3.	Urea alone 110g/plant	117.0	50.0	154.8	

Subramaniyan S. and Selvaraj (1986)

Grapevine: The root knot nematode *Meloidogyne incognita* is prevalent in vine yards of South India and is one of the limiting factors in grapevine cultivation. Varieties like Anab-e-shahi, Hamlat, Muscat humburg and Arkavathi lodges a number of phytonematodes. Species of *Xiphinema, Longidorus, Helicotylenchus* and *Hoplolaimus* were found to occur concomitantly with *M. incognita* in grapevine gardens.

Since, *M. incognita* poses to be a problem in vine cultivation, establishment, growth and stand of vines gets affected due to the nematode. The leaf size is reduced and cupping of leaves occur in addition to chlorosis of leaves and premature leaf fall, ultimately it reflects on the yield. Experiments were conducted to evaluate botanicals like neem in comparison with chemicals for nematode management.

Two experiments were conducted in 2 seasons for managing the nematodes. Neem cake and castor cake were tried individually and in combination with chemicals as also with urea coating in comparison with individual chemicals and controls.

The results appended in Table 3 revealed that all the treatments were found superior over control. The number of galls per g. of root was lowest with carbofuran 3G, @ 2g. a.i/vine followed by Neem cake 0.025N + Carbofuran 1g. a.i/vine

(32.91 percent reduction over control). Neem coated urea 0.1N + carbofuran 1g a.i/ha/vine registered 28.39 per cent reduction. Comparing the yield of fruits the vines treated with neem cake 0.025N + carbofuran 1g a.i/vine recorded highest yield of 129.39 bunches/plot (13.415kg) which amounts to 14.76 tonnes/ha. as against 94.67 bunches (8.185 kg) and 9 tonnes/ha, which is in the tune of 64.0 per cent over untreated control. Neem cake alone @ 0.05 N/vine has registered 35.4 per cent increased yield over control (Table 4)

Table 3. Nematicidal Trial effect of treatments on gall density in grapes CV Muscat Hamburg (I. season-Aug, Dec 1985)

Sl. No.	Treatments	Galls per gram of roots (Mean of 3 replications)					Mean	Per cent reduction over control
		Before treatment	30th day	50th day	75th day	At harvest		
1.	Phorate 2g a.i./vine	63.7	40.7	45.0	57.3	76.0	56.5	27.93
2.	Carbofuran 2g a.i./vine	63.0	37.0	41.3	47.0	73.0	55.3	35.88
3.	Quinalphos 2g a.i./vine	65.7	36.0	42.3	55.3	79.7	55.8	28.82
4.	Neem cake 0.05 N/vine	61.0	53.7	56.3	62.3	69.0	60.5	22.90
5.	Castor cake 0.05 N/vine	63.3	41.0	47.0	53.3	64.7	53.9	31.29
6.	Neem cake 0,025 N + Phorate 1g a.i./vine	63.3	43.7	48.7	66.0	74.7	59.5	24.15
7.	Neem cake 0.025 N + Carbofuran 1g a.i./vine	60.0	39.3	47.0	51.7	65.0	52.6	32.91
8.	Neem cake 0.025 N + Quinalphos 1g a.i./vine	65.0	47.0	53.0	58.7	74.0	95.5	24.07
9.	Castor cake 0.025 N + Phorate 1g a.i./vine	69.0	48.7	53.0	65.3	72.3	61.8	21.17
10.	Castor cake 0.025 N+Carbofuran 1g a.i./vine	69.3	43.7	51.0	58.0	64.0	57.2	27.04
11.	Castor cake 0.025 N+Quinalphos 1g a.i./vine	64.0	41.7	46.7	57.3	70.7	56.1	28.48
12.	Neem coated urea 0.2 N/vine	68.3	43.7	50.7	56.0	74.3	58.6	25.26
13.	Neem coated urea 0.1 N + phorate 1g a.i./vine	73.3	55.3	61.0	71.3	77.3	56.1	28.39
14.	Neem coated urea 0.1 N + Carbofuran 1g a.i./vine	59.7	42.7	49.7	55.0	73.7	56.1	28.39
15.	Neem coated urea 0.1 N + Quinalphos 1g a.i./vine	63.3	44.3	49.3	55.3	76.3	57.7	26.36
16.	Control	69.0	79.0	88.0	80.3	75.7	78.4	—
	Mean	65.1	46.1	52.0	59.4	71.9		

Nathan, A.J. (1986)

CD (P + 0.05) i) Between treatments : 6.60 ii) Between periods : 3.68 iii) Interaction between periods and treatments : N.S.

In another set of experiment, two levels of neem cake (0.05 N/vine and 0.10 N/vine), two levels of carbofuran (2g and 3g ai/vine) and combination of half the dose of neem cake and half the dose of carbofuran were tried. The results indicated that all the treatments were found superior over control. Among the treatments carbofuran @ 3g ai/vine recorded lowest incidence of galls (40.8 galls/g root)

Table 4. Nematicidal effect on the yield of grapevine (I season)

Sl. No.	Treatments	Bunches/ plot	Yield kg/plot	Calculated yield t/ha	Percentage of increase over control
1.	Phorate 2g a.i./vine	105.67	9.935	10.93	21.4
2.	Carbofuran 2g a.i./vine	116.67	11.685	12.85	42.3
3.	Quinalphos 2g a.i./vine	106.00	9.435	10.38	15.3
4.	Neem cake 0.05 N/vine	107.67	11.085	12.19	30.4
5.	Castor cake 0.05 N/vine	118.00	11.58541.5		12.74
6.	Neem cake 0,025 N + Phorate 1g a.i./vine	115.67	10.815.	11.90	32.2
7.	Neem cake 0.025 N + Carbofuran 1g a.i./vine	129.39.	19.415	14.76	64.0
8.	Neem cake 0.025 N + Quinalphos 1g a.i./vine	171.00	10.165	11.18	24.2
9.	Castor cake 0.025 N + Phorate 1g a.i./vine	108.33	11.715	12.39	43.2
10.	Castor cake 0.025 N + Carbofuran 1g a.i./vine	125.67	12.500	13.75	52.8
11.	Castor cake 0.025 N + Quinalphos 1g a.i./vine	115.67	10.350	11.38	26.4
12.	Neem coated urea 0.2 N/vine	113/67	9.835	10.82	20.2
13.	Neem coated urea 0.1 N + phorate 1g a.i./vine	116.67	9.665	10.63	18.1
14.	Neem coated urea 0.1 N + Carbofuran 1g a.i./vine	111.67	10.200	11.22	24.7
15.	Neem coated urea 0.1 N + Quinalphos 1g a.i./vine	113.00	9.235	10.16	12.9
16.	Control	94.67	8.125	9.00	
	C.D. (P = 0.05)	NS	0.71		

(A.J. Mathan 1986)

followed by neem cake 0.10 N/vine (46.1 galls/g root) and neem cake 0.05 N + Carbofuran 1.5g a.i/vine (47.7 galls/g. root) as against 81.4 galls in the untreated control (Table 5)

Table 5. Nematicidal - effect on gall density of *M. incognita* of grapevine (II season Jan-May, 1986)

Sl. No.	Treatments	Mean of 3 replications Galls per gram of roots					Mean	Percent reduction control
		Before treatment	30th day	50th day	75th day	At harvest		
1.	Neem cake 0.05 N/vine	71.0	59.3	31.3	53.3	62.7	55.5	31.78
2.	Neem cake 0.10 N/vine	73.3	38.0	20.9	44.7	54.7	46.1	43.33
3.	Carbofuran 2g a.i./vine	67.3	50.3	31.7	53.7	60.0	52.6	35.38
4.	Carbofuran 3g a.i./vine	65.7	29.3	19.7	39.3	50.0	40.8	49.88
5.	Neem cake 0.025 N + Carbofuran 1g a.i./vine	74.7	56.3	31.3	60.7	69.3	58.5	28.17
6.	Neem cake 0.025 N + Carbofuran 1.5g a.i./vine	66.7	52.3	33.7	53.3	61.3	53.5	34.31
7.	Neem cake 0.05 N + Carbofuran 1g a.i./vine	69.3	57.0	49.3	52.7	56.7	57.0	29.98
8.	Neem cake 0.05 N + Carbofuran 1.5g a.i./vine	68.0	47.7	35.0	37.0	51.0	47.7	41.36
9.	Control	66.7	84.7	94.0	35.7	76.0	81.4	—
	Mean	69.2	52.8	38.4	53.4	60.2		

CD (P = 0.05) i) Between treatments : 3.62 ii) Between periods : 4.87 iii) Interaction between periods and treatments : 10.87

(A.J. Mathan, 1986)

The yield data appended in Table 6, revealed that the vines applied with neem cake 0.05 N + carbofuran 1.5g a.i/vine registered highest fruit yield of 15.18 tonnes/ ha as against 10.50 t/ha. The treatment has increased the fruit yield to the tune of 44.57 percent.

Table 6. Nematicidal - effect on the yield of Grapevine (II season - Jan-May, 1986)

Sl. No.	Treatments	Bunches/ plot	Yeild (kg/plot)	Calculated yield t/ha	Percentage of increase in yield over control
1.	Neem cake 0.05 N/vine	116.67	10.900	14.19	11.99
2.	Neem cake 0.10 N/vine	126.67	12.300	13.53	28.86
3.	Carbofuran 2g a.i./vine	133.00	11.400	12.54	19.43
4.	Carbofuran 3g a.i./vine	137.33	12.400	13.75	30.95
5.	Neem cake 0.025 N + Carbofuran 1g a.i./vine	121.67	12.300	13.53	28.86
6.	Neem cake 0.025 N + Carbofuran 1.5g a.i./vine	120.65	12.500	13.75	30.95
7.	Neem cake 0.05 N + Carbofuran 1g a.i./vine	135.33	13.400	14.74	40.38
8.	Neem cake 0.05 N + Carbofuran 1.5g a.i./vine	132.33	13.000	15.18	44.57
9.	Control	105.00	9.550	10.50	—
	CD (P = 0.05)	16.45	1.09		

(A.J. Mathan, 1986)

Attempts were made to manage the root-knot nematode, *M. incognita* under inoculated condition on the cv. Arkavathi. Neem cake alone (250/vine) and in combination with carbofuran (60g/vine) and phorate (15g/vine) were tried. Neem cake was also tried with half the dose of carbofuran and phorate. The observations on the length and width of shoot, number of branches, number of leaves, root length and weight, appended in Table 7 revealed that neem cake 250g/vine + half the dose of carbofuran registered increased growth parameters like shoot and root. The treatment registered 129.45 per cent increase in the number of branches over untreated control. The number of galls/g. root was lowest (2.1) with carbofuran followed by neem cake 250g with half the dose of carbofuran (3.6) as against 45.1 in the untreated control. Neem cake + 1/2 dose of carbofuran registered 2.8 egg masses per g root as against 20.2 in the inoculated control (Table 8).

Table 7. Management of *M. incognita* on grapevine cv. 'Arkavati' (Jan-May, 1993)

(mean of four)

Sl. No.		Treatments	Shoot		No. of	Leaves		Root	
			Length (cm)	Weight (gm)	branches	Number of leaves	Leaves weight	Length (cm)	Weight (gm)
1.	T_1	Neem cake alone (NC) (250 g/vine)	249.24	127.60	5.26	141.92	111.32	58.26	100.85
2.	T_2	NC (250 g/vine) + 1/2 D of Carbofuran (30 g/vine	302.53	157.50	6.31	170.03	131.07	62.88	94.29
3.	T_3	NC (250 g/vine) + 1/2 D of phorate (7.5 g/vine)	277.73	141.56	5.82	156.57	121.25	60.14	92.26
4.	T_4	Carbofuran (60 g/vine)	217.50	133.26	4.54	122.35	96.34	75.54	81.78
5.	T_5	Phorate (15 g/vine)	210.50	110.54	4.38	120.54	95.05	67.91	86.53
6.	T_6	Uninoculated control	221.50	112.06	4.64	123.50	98.24	66.57	89.05
7.	T_7	CD at 5%	127.25	66.70	2.75	70.06	50.79	38.32	111.56
		CD at 5%	21.40	8.59	0.57	12.53	8.23	6.78	7.82

(J. Sankarganesh, 1993)

Table 8. Management of root knot nematode *M. incognita* on its final population and gall formation

Sl. No.	Nematodes/kg of soil	Final Population 250 cc soil	No. of galls/gm of root	No. of egg masses/g root
1. T_1	Neem cake (250 g/vine)	37.20.6.8	5.7	
2. T_2	NC (250 g/vine) + 1/2 D of Carbofuran (30 g/vine	21.65	3.6	2.8
3. T_3	NC (250 g/vine) + 1/2 D of phorate (7.5 g/vine)	29.31	3.8	3.7
4. T_4	Carbofuran (60 g/vine)	15.32	2.1	1.5
5. T_5	Phorate (15 g/vine)	18.19	2.5	2.2
6. T_6	Uninoculated control	0	0	0
7. T_7	CD at 5%	126.3	45.1	20.2
	CD at 5%	1.23	0.43	0.37

(J. Sankarganesh, 1993)

Tomato: The root knot nematode, *M. incognita* is a major problem in tomato. A field experiment was conducted with neem cake, mahua cake and compost @ 500 kg/ha, and ethoprop @ 1kg a.i./ha. Combination of half the dose of the amendments were also tried with ethoprop. The minimum root knot index of 3.0 was recorded in treatment neem cake 250 kg + ethoprop 0.5 kg ai/ha as against 5.0 in the control. The percent reduction in the nematode population was also more highest (75.18 per cent) in the plots treated with neem cake @ 250 kg + ethoprop @ 0.5 kg a.i/ha followed by compost @ 250 kg + ethoprop @ 0.5 kg ai/ha, (Table 9). Application of neem cake @ 250 kgs/ha + ethoprop @ 0.5 kg ai/ha. resulted in highest phenol accumulation in the roots (1.050 µg/g) as followed by application of neem cake alone @ 500 kg/ha as against 0.450 µg in the untreated check (Table 10).

Table 9. Effect of organic amendments and nematicide application on the incidence of *M. incognita* in tomato cv. Marudham

	Treatments	Root knot index		Soil population
		Mean	% of decrease over control	(% decrease/increase over initial inoculum)
T_1	Compost @ 500 kg/ha	4.00	-10.55	-56.60
T_2	Mahua cake @ 500 kg/ha	4.00	-10.55	-61.07
T_3	Neem cake @ 500 kg/ha	4.00	-10.55	-63.75
T_4	Ethoprop @ 1 kg ai/ha	3.61	-15.02	-66.81
T_5	Compost @ 250 kg + Neem cake @ 250 kg/ha	4.00	-10.55	-67.59
T_6	Compost @ 250 kg + Mahua cake @ 250 kg/ha	3.87	-12.03	67.30
T_7	Compost @ 250 kg + Ethoprop @ 0.5 kg ai/ha	3.61	-15.02	-70.05
T_8	Neem cake @ 250 kg + Ethoprop @ 0.5 kg ai/ha	3.00	-22.59	-75.18
T_9	Mahua cake @ 250 kg + Ethoprop @ 0.5 kg ai/ha	3.49	-16.45	-67.72
T_{10}	Control (untreated check)	5.00	+30.47	
	Sed	0.051	1.5267	
	CD (P = 0.05)	3.1329	3.1329	

(Gulsar Banu, 1992)

Table 10. Effect of organic amendments and nematicide application on the phenolic content of tomato root

	Treatments	Total phenols (ug/g of root) dry wt. basis
T_1	Compost @ 500 kg/ha	0.500
T_2	Mahua cake @ 500 kg/ha	0.825
T_3	Neem cake @ 500 kg/ha	0.900
T_4	Ethoprop @ ! kg ai/ha	0.775
T_5	Compost @ 250 kg/ha + Neem cake @ 250 kg/ha	0.850
T_6	Compost @ 250 kg/ha + Mahua cake @ 250 kg/ha	0.693
T_7	Compost @ 250 kg/ha + Ethoprop @ 0.5 kg ai/ha	0.700
T_8	Neem cake @ 250 kgs/ha + Ethoprop @ 0.5 kg ai/ha	1.050
T_9	Mahua cake @ 250 kgs/ha + Ethoprop @ 0.5 kg ai/ha	0.950
T_{10}	Untreated check	0.450
	SE	0.1893

(Gulsar Banu, 1992)

Tomato seeds (cv. Co3) were soaked in margocide CK 20 EC, neem seed kernal extract 5%, neem oil 0.1% EC, neem cake extract 5% and with insecticides viz. phosphomidon 85 WSC 0.1%, carbosulfan 25 ST 0.1%, dimethoate 30EC 0.1% and sown.

The results appended in Table 11 revealed that the seeds soaked with phosphomidon 0.1% has increased the shoot height by 200% followed by dimethoate 0.1% (193.*%) and margocide (K 20EC @ 0.1% (188.8%) (Table i1).

Table 11. Effect of seed soaking on root-knot nematode, *Meloidogyne incognita* on tomato (Co 3) a - Plant growth parameters (Mean of 3) (1992)

S. No.	Treat-ments	Shoot Height (cm)	% increase/ decrease	Weight (g)	% increase/ decrease	Root Lenght (cm)	% increase/ decrease	Weight (g)	% increase/ decrease
1.	T_1	15.23	+85.0	6.16	+188.8	6.67	+49.2	1.40	+75.00
2.	T_2	9.73	+18.2	3.60	+68.8	6.30	+40.9	1.00	+25.00
3.	T_3	14.10	+71.3	5.27	+146.9	6.76	+51.2	1.20	+50.00
4.	T_4	12.97	+57.5	4.10	+92.2	7.00	+56.6	1.15	+43.75
5.	T_5	20.50	+148.9	6.40	+200.0	11.00	+146.1	1.63	+104.10
6.	T_6	16.50	+100.4	5.83	+73.5	7.97	+78.3	1.53	+91.60
7.	T_7	18.43	+123.9	6.27	+193.8	8.57	+91.7	2.32	+190.00
8.	T_8	8.23	2.13	4.47	0.80				
CD(P=0.05)		0.36		0.13		0.42		0.24	

T_1 = Margocide CK 20EC 0.1% T_2 = *A.indica* seed kernel extract 5% T_3 = Neem oil 50 EC 0.1%
T_4 = *A.indica* cake extract 5% T_5 = Phosphomidan 85 WSC 0.1% T_6 = Carbosulfan 25 ST 0.1%
T_7 = Dimethoate 30 EC 0.1% T_8 = Plain water

(Gulsar Banu 1992)

The seedlings of tomato (Co 30 was subjected to bare root dip in the neem products viz. margocide CK 20 EC 0.1%, neem seed kernal extract 5%, neem oil 0.1%, neem cake extract 5%, phosphomidon 85 WSC 0.1% and dimethoate 30EC 0.1% and transplanted in the pots. The observations on the root knot index, Number of females per g. root, number of egg masses per g root and soil population of nematodes were assessed.

The results appended in Table 12 revealed that the lowest root knot index (1.670 was observed with dimethoate root dip, followed by margocide CK 20 EC 0.1% (2.33). Margocide root dip has registered 79.0 percent decrease in number of females, 95.0 per cent decrease in the number of egg masses/g. root, 88.4 per cent reduction in the number of eggs per egg mass, and 81.4 percent reduction in the soil population of *M. incognita*. However, this treatment was on par with dimethoate 30 EC 0.1%.

Table 12. Effect of seedling bare-root dip on root-knot nematode, *Meloidogyne incognita* on tomato (Co 3) b. Nematode populaation (mean of 3) (1772)

Sl. No.	Treatments	Root knot index	Number of females/ g root	% increase/ decrease	Number of egg masses/g root	% increase/ decrease	Number of eggs/ egg mass	% increase/ decrease	Population/ 100 cc soil	% increase/ decrease
1.	T_1	2.33	20.67	-79.0	3.67	-95.3	21.67	-88.4	228.30	-81.4
2.	T_2	3.33	62.33	-36.6	22.33	-71.5	104.67	-43.9	284.00	76.8
3	T_3	2.67	25.33	-74.2	10.00	-87.2	74.67	-60.0	254.00	-79.3
4.	T_4	3.00	40.38	-58.9	16.00	-79.6	92.33	-50.5	296.60	-75.8
5.	T_5	2.33	22.00	-77.0	8.33	-89.4	55.00	-70.5	211.00	-82.8
6.	T_6	1.67	15.67	-84.1	7.00	-91.1	29.67	-84.1	185.60	-84.4
7.	T_7	5.00	98.33		78.33		186.67		1225.00	
CD(P=0.05)		0.91	5.71		3.44		5.99			

T_1 = Margocide CK 20 EC 0.1% T_2 = *A.indica* seed kernel extract 5% T_3 = Neem oil 50EC 0.1%
T_4 = *A.indica* cake extract 5% T_5 = Phosphomidan 85WSC 0.1% T_6 = Dimethoate 30 EC 0.1%
T_7 = Plain water
(Gulsar Banu, 1992)

In respect of the growth parameters margocide CK 20 EC was highly effective which recorded 369.6 per cent increase in plant height, 255.0 per cent in shoot weight, 135.6 per cent increase in root length and 135.5 per cent increase in root weight. The treatments, viz., neem kernal extract 5% and neem oil 50 EC (0.1%) were on par (Table 13)

Betelvine (*Piper betle*.L.): Betelvine is one of the cash crops of Tamil Nadu affected by a number of nematodes, like *Meloidogyne incognita, Rotylenchulus reniformis, Helicotylenchus incisus* etc.

Field experiments were conducted with de-oiled cakes of neem, groundnut, coconut, castor, poultry litter, saw dust and calotropis leaves. The oil cakes were applied

Table 13. Effect of seedling bare root dip on root-knot nematode *Meloidogyne incognita* plant growth parameters (mean of 3) (1992)

S. No.	Treat-ments	Shoot				Root			
		Height (cm)	% increase/ decrease	Weight (g)	% increase/ decrease	Lenght (cm)	% increase/ decrease	Weight (g)	% increase/ decrease
1.	T_1	27.23	+369.5	4.73	+255.9	5.73	+135.6	2.43	+135.5
2.	T_2	16.07	+177.0	2.62	+96.7	5.40	+121.9	1.47	41.9
3	T_3	16.20	+179.3	2.23	+67.9	5.33	+119.1	1.50	+45.2
4.	T_4	12.53	+116.1	1.70	+27.8	4.90	+101.4	1.20	+16.2
5.	T_5	17.33	+198.8	2.63	+98.0	5.93	+143.9	1.50	+45.2
6.	T_6	21.90	+277.6	4.13	+210.8	5.40	+121.9	2.03	+96.8
7.	T_7	5.80		1.33		2.43		1.03	
CD(P=0.05)			0.34		0.65		0.40		

T_1 = Margocide CK 20 EC 0.1% T_2 = *A.indica* seed kernel extract 5% T_3 = Neem oil 50EC 0.1%
T_4 = *A.indica* cake extract 5% T_5 = Phosphomidan 85WSC 0.1% T_6 = Dimethoate 30 EC 0.1%
T_7 = Plain water
(Gulsar Banu, 1992)

@ 100N/ha and saw dust and calotropis leaves were applied @ 2.5 tonnes/ha. The nematode population reduction and leaf yields were recorded. The data (Table 14) indicate that calotropis leaf application @ 2.5 t/ha has reduced the nematode population by 59.7 per cent followed by neem cake @ 100 N/ha with 44.4 percent. The highest leaf yield was also recorded in calotropis leaf applied beds (14.2kg and 4800 leaves) followed by neem cake (4200 leaves & 12.1kg)

The treatment also reduced the incidence of betelvine wilt caused by *Phytopthora* sp. (Table 14).

Table 14. Betlevine (Karpoori) - organic amendments on nematode population, leaf yield and wilt incidence - 1984 - 1987 - (3 replications)

Sl. No.	Treatments	Nematode population (% change)	Leaf yield Wt (kg)	Leaf yield No. (000s)	Wilt incidence (%)
			(3 Years mean)		
1.	Neem cake 100 N/ha	-44.4	12.1[c]	4.2[b]	10.8[a]
2.	Groundnut cake 100 N/ha	-22.9	11.0[b]	3.8[a]	34.4[c]
3.	Castor cake 100 N/ha	-25.9	11.0[b]	4.0[ab]	34.4[c]
4.	Coconut cake 100 N/ha	-28.7	11.4[b]	3.8[a]	29.8[b]
5.	Poultry litter 100 N/ha	-40.9	12.3[c]	4.2[b]	17.5[a]
6.	Sawdust 2.5 t/ha	-29.3	11.3[b]	3.9[a]	17.0[a]
7.	Calotropis leaves 2.5 t/ha	-59.7	14.2[d]	4.8[b]	8.1[a]
8.	Control	+166.7	9.4[a]	3.0[a]	49.7[d]
	C.D. (P = 0.05)		0.5	0.98	10.21

(Sivakumar *et al.* 1987)

In the second set of experiment, sawdust 2.0 t/ha, neem cake 1.0 ton/ha and carbofuran 3G @ 1.5 kg a.i/ha were tried. Half the dose of amendments were also tried with half the dose of carbofuran. The data furnished in Table 15 shows that highest reduction of nematode population (43.45) was brought about by application of carbofuran 3G @ 1.5 kg a.i/ha, followed by neem cake 0.5 t/ha + carbofuran 0.75 kg a.i/ha (41.32% reduction). The highest leaf yield (4120 leaves) was recorded in beds treated with neem cake 0.5 t/ha + carbofuran 0.75 kg a.i/ha.

Table 15. Betelvine (Karpoori) - organic amendments and systemic granules on nematode and leaf yield - 1985 -1987 (4 replications)

Sl. No.	Treatments	Nematode Population change	Leaf yield (Nos.)
1.	Sawdust 2.0 t/ha	-25.73	2703[a]
2.	Neem cake 1.0 t/ha	-27.91	3052[b]
3.	Carbofuran 1.5 kg a.i/ha	-43.45	2792[ab]
4.	Sawdust 1.0 t/ha + Carbofuran 0.75 kg a.i/ha	-27.91	2292[a]
5.	Neem cake 0.5 t/ha + Carbofuran 0.75 kg a.i/ha	-41.32	4120[c]
6.	Control	+229.03	1943[a]
	C.D. at 5%		784

(Sivakumar *et al.* 1987)

In the third experiment, neem cake was applied at three levels viz. 0.5, 1.0 and 2.0 t/ha. Sawdust at two levels viz. 1.0 and 2.0 tons/ha with and without NPK at 150:100:50, and NPK alone at 150:100:50 were tried. Half the dose of the nitrogen was supplemented by neem cake and the rest through urea.

The results indicated that the highest reduction in the nematode population was noticed (51.10%) in the beds applied with neem cake at 2.0 tons/ha, whereas, the highest yield of betel leaf (3638) was recorded in the beds applied with sawdust 2.0 tons/ha + NPK at 75+75 100:50 kg/ha in which half the dose of N was supplemented through neem cake and the rest through urea. Neem cake application at 2.0 ton/ha registered 3432 leaves/bed.

Table 16. Betelvine (Karpoori) - organic amendments and nematode population and yield - 1986 - 1987 (one year old -3 replications)

Sl. No.	Treatments	Location - I		Location - II	
		Nematode population change	Leaf yield ('00s)	Nematode population change	Leaf yield ('00s)
1.	Neem cake 0.5 t/ha	-15.95	24.54[a]	-17.18	25.55[a]
2.	Neem cake 1.0 t/ha	-22.29	29.56[b]	-29.05	27.70[b]
3	Neem cake 2.0 t/ha	-51.10	34.32[c]	-27.01	32.36[d]
4.	Sawdust 1.0 t/ha	-9.10	23.96[a]	-21.60	25.44[b]
5.	Sawdust 2.0 t/ha	-21.77	23.47[a]	-13.99	34.59[b]
6.	Sawdust 1.0 t/ha + NPK 75+75 : 100 : 50	-13.31	23.58[a]	-15.28	29.42[c]
7.	Sawdust 2.0 t/ha + NPK 75+75 : 100 : 50	-16.09	36.38[c]	-17.30	28.53[bc]
8.	NPK alone 75 : 100 : 50	-12.28	28.20[ab]	-4.18	27.29[b]
9.	Control	+44.68	26.08[a]	-69.87	19.78[a]
	C.D. 1%	3.30		2.42	

(Sivakumar *et al.* 1987)

REFERENCES

Gulsarbanu, J. 1992. Efficacy of botanicals against *Meloidogyne incognita* (Kofoid and White, 1919) Chitwood, 1949. on tomato *Lycopersicon esculentum* Mill). M.Sc.(Ag.) Thesis submitted to the Tamil Nadu Agricultural University, Coimbatore. India.

Mathan, A.J. 1986. Studies on the management of root knot nematode, Berry thrips and mildews on grapevine (*Vitis vinifera* L) M.Sc.(Ag.) Thesis submitted to the Tamil Nadu Agricultural University, Coimbatore, India.

Sankarganesh, J. 1993. Population Studies of phytonematodes and management of *Meloidogyne incognita* (Kofoid and White 1919) Chitwood 1949. in grapevine (*Vitis vinifera* L). M.Sc.(Ag.) Thesis submitted to the Tamil Nadu Agricultural University, Coimbatore. India.

Sivagami Vadivelu, R., S. Subramanian, M. Sivakumar and S. Jayaraj. 1988. Botanicals for nematode control in rice based crops in Tamil Nadu. Presented at the Final Workshop at IRRI-ADB-EWC. Project on Botanical Pest Control in Rice-based cropping systems. IRRI, Los Banos., Laguna, Philippines.

Sivakumar, M., Natarajan, S. and M. Balasubramanian, 1987. Nematode pests of betelvine (*Piper betle* L.) and their management. Paper presented at the third nematological group meeting on Nematode pests of plantation crops. Sugarcane Breeding Institute, Coimbatore. November 24-27.

Subramaniyan, S and P. Selveraj 1986. Annual Report. Banana Research Station, TNAU Vellore.

Neem for the Management of Crop Diseases, 1995
Editor : V. Mariappan
Associated Publishing Co., New Delhi, India
pp. 201-207.

SIMULTANEOUS CONTROL OF NEMATODES AND ROOT ROT FUNGAL PATHOGENS OF SOYBEAN BY NEEM BASED PRODUCTS

Vyas, S.C.

Department of Plant Pathology, J.N. Agricultural University
Indore-452 001, M.P.

INTRODUCTION

The crops suffer from more than one pathogens that may be fungi or bacteria or nematodes. These pathogens attack the crop simultaneously and accordingly the crop is treated with biocides toxic to individual pathogens. These resulted is high production cost and labour. The agricultural scientists particularly the plant pathologists and entomologists are searching such compounds which can control both the pathogens and pests. Neem cake and oil extracted from tree. *Azadiracta indica*, possesses such properties. Mammen (1972), Devkumar (1985) and Kathirvel *et al.* (1992) reported that soil and seed application of neem extract control root knot nematodes (RKN) population and infection in several crops. Neem contains *azadiractin* which is highly toxic and lethal to the larvae of *Meloidogyne* spp. and also it inhibits hatching of eggs (Hussain and Maszood, 1976; Bhatnagar *et al.* 1980). Seed treatment with neem leaf extract at the rate of 0.3 kg per three litre of water and *mahua* oil reduced the invasion of *Bhindi* roots by RKN to an appreciable extent (Kathirvel *et al.* 1992). Egunjobi and Afolani (1976) also reported effective control of *Pratylechus brachyrus* and growth and yield increase in maize by neem cake.

This study was designed to evaluate the effects of neem cake and oil and Repellin on the root rot and root knot complex of soynean which is cultivation problem in the Malwa and adjoining areas and to evaluate the effect on the biofertilizing agents such as mycorrhiza and rhizobia and subsequently the yield of the soybean which is of a practical significance from the cultivator points of view.

MATERIALS AND METHODS

Soybean seeds of cultivar, JS-72-44, a highly susceptible variety to various pests and diseases was used in these experiments. Recommended practice of fertilizer 20 kg nitrogen and 40 kg superphosphate per hactare was applied. Seeds of soynean were treated with thiram at 3g per kg and subsequently *Rhozobium japanicum* was applied at the time of sowing at the rate of 10 g per kg. Soil was infested with mycorrhizal culture comprising root pieces predominantly of *Glomus* spp. at the rate 100 kg root pieces per hactare. The compatibility between fungicide, rhizobium and mycorrhizae was already established (Vyas, 1993) in earlier studies. The soil was infested with *Rhizoctonia bataticola* and *Scerotium rolfsii* cultures a week before sowing and tested that root rot and collar rots appear in severe forms constantly. Application of neem cake, neem oil, Repellin and carbofuran was carried out after the soil infestation with the fungal cultures. In case of neem oil and neem cake obtained as commercial formulation. All these ingredients were applied at the rate of 1 and 2%.

The mixed inoculum of nematodes comprising of several species was obtained from a glass house on solanaceous crops and which was pathogenic to the soybean also. Composite samples of adopting combined "Gravity and Baermann funnel method (GBFM) for soil (Christie and Perry, 1952) and "Blending and Seivinf method" for quantifying and qualifying the various genera associated with soybean was carried out by the key developed by Mail and Lyon (1975) and Taylor (1967).

The observations for the root nodules and weights were recorded after 21 DAS, mycorrhizal infection after 65 DAS and germination after 21 DAS and yield after the harvest as per methods developed by Vyas (1993).

RESULTS AND DISCUSSION

Suppression of economically important nematodes and soilborne through soil amendments with carbonaceous sources of organic matter is well documented. Also equally important is integrating other methods of plant disease control. The present study was undertaken with the object that neem extract and oil and neem based preparation Repellin possess potential to control root rot pathogens and nematodes which has been amply elucidated in the literature (Mammen, 1972; Kathirvel *et al.*, 1992 and others).

Soybean suffers from several plant pathogens and nematodes. During the surveys of soybean fields in university farm and also adjoining cultivators farms where soybean was cultivated since last twenty years, several nematodes are associated in different fields. The perusal of data in Table-1 indicate that in soil samples from

Table 1. Average density and frequency of occurrence of different plant parasitic nematode species

Soil	Density and frequency of nematodes in 250 cc soil/5g root						
1	2	3	4	5	6	7	
Sample 1	700	100	175	300	200	500	1975
Sample 2	500	150	1150	300	150	700	1950
Sample 3	300	200	200	300	150	600	1750

1 - *Helicotylenchus multicinctus* 2 - *Xiphenema Index* 3 - *Pratylenchus Coffee* 4 - *Aphelenchus Sp.* 5 - *Heterodere glycine and* 6 - *Meloidogyne incognita*

soybean cropped soil contain *Helicotylenchus multicinctus Meloidogyne incognito* and *Aphelenchus Sp.* occur in maximum number as compared to *Xiphenema index, Pratylenchus coffee* and *Heterodera glycine.Heterodera glycine* is reported to be pathogenic to soybean United States (Taylor, 1967). A considerable number of nematode unidentified were also encountered which neemd further identification. Pathogenicity that was also carried out by mixed inoculum and all most all the prevailing varieties such as JS.72-44, PK-474, Punjab-1 and JS-75-05 were found susceptible to highly susceptible. Soybean variety JS-72-44 was found highly susceptible and taken for experimental works. The idea of giving this table-2 is that the soil contains different nematodes which incite different symptoms which cause wide confusion.

Nematode damage indicates a varieties of symptoms. Although it is dificult to single out symptom produced by different nematode species associated with soybean because of overlapping and borderline differences, however, based on several reports, the distinguishing symptoms of the various species are listed in the Table-2. By and large, the aerial symptoms comprise of stunting, dwarfing, chlorosis, yellowing, necrosis and various types of leaf spotting and mottling associated with ground symptoms such as root lesions and root knots. Nematode damage will likely show up as circular areas in the infested fields with stunted yellow plants at the centre grading out to green normal infected by nematodes are usually conspicuous because of the reduced unthrifty growth and a tendency to wilt on warm days. Heavy infections result in a severe loss of yield, reduced quality of produce accompanied by deficiency symptoms. It was observed that nematode infection was associated with increased susceptibility to foliage and soilborne diseases such as collar and dry root rots caused by *Sclerotium rolfsii* and *Rhizoctonia bataticola, respectively.*

Table 2. Effect of several species of nematodes on the occurrence of plant symptoms in soybean sp. in artificially inoculated soil

Aphelenchus sp	*Helicotylenchus multicinctus*	*Pratylenchus Penetran*	*Xiphinema index*	*Heterodera slycine*
Chlorasis, dwarfing, stunting and sickness	Nematode causes symptoms in cortex of roots and easily introduced in fresh area with infested soil and plant tissues.	Nematode is associated with soil sickness and names root lesion nematodes; they are parasites of corex but may damage vascular tissues in soybean.	Migratory root ectoparasite; they also transmit viruses in grapes. They cause necrosis, terminal swelling, celluar hypectrophy & multinucleate condition of cortical cells near the feeding site.	Disease plants with nematode start yellowing as early as one month after planting; after carefully removing soil from roots reveal creamy coloured cysts.

The results presented in Table-3 indicate that the soybean suffers from nematodes (Table-1 and 2) and also soilborne fungal pathogens infection such as collar rot (*Sclerotium rolfsii*) and dry root rot (*Rhizoctonia bataticola*) to that extent that economic cultivation of this crop will not be possible in the foreseeable future. The experiments conducted under artificial conditions that is by creating disease and infection of nematodes the application of neem cake, neem oil, Repellin and carbofurn reduced the infection and diseases at significant level and also increased the yield significantly. Among the toxicants, Repellin was found highly significant in reducing the nematode population and disease intensities. The toxicity descending order to nematodes was Repellin, neem oil, carbofuran and neem cake but all the treatment were significantly superior over the control. The application of neem based products were equally toxic to the nematodes and fungi which is amply evident from the data presented in the Table-2. Similar results have been reported for RKN by Mammen (1972), Devkumar (1985) and Kathirvel *et al.* (1992) and Murthy and Rao (1992). These results suggest that application carbonecious organic matter to the soil eradicate pathogenic soil biota significantly and also ameolarate the soil for better crop growth.

Table 3. Effect of soil application of neem cake, neem oil, Repellin and carbofuran at the rate of 1 and 2 percent of soil weight on germination percentage of soybean, nematode population, Mycorrhizal infection and nodule number and weight of soybean (cultivar - JS-72-44)

Treatment	Conc. (a.i. in %)	Germination (%)	*Nematode* Population actual count	Population reduction (% over control)	Mycorrhiizal infection	Root rot (%)
Neem cake	1.00	70 b	1500	62.50 c	68.00 b	10.00 c
	2.00	75 a	1300	67.50 c	70.00 b	09.00 c
Neem oil	1.00	60 c	1250	50.00 d	65.00 b	12.00 b
	2.00	65 b	1200	68.25 c	65.00 b	12.00 b
Repellin	1.00	70 b	0850	80.00 b	75.00 a	07.00 a
	2.00	75 a	0250	93.25 a	75.00 a	09.00 a
Carbofuran	1.00	65 b	1600	60.00 c	59.00 c	15.00 b
	2.00	65 b	1300	67.25 b	60.00 c	15.00 b
Check	—	35 d	4000	—	55.00 d	20.00 d

* Data are mean of three replications

** Seed treatment at the rate of 0.2 mg a.i. per kg was given with thiram and soil inoculation with mycorrhizal fungi and seed inoculation with *Rhizobium japanicum*; before sowing;

*** Number not followed by the same letter are significantly different at the P=0.05 in Duncan's multiple range tests;

**** Soil was inoculated with *Sclerotium rolfsii*, and *Rhizoctonia bataticola*before application of the treatment

***** Disease index after 40 days after sowing

With the world fuel shortage, high cost and nonavailability of agrochemicals, application becomes a problem and world over agriculturists, farmers and administrators are worried. A sort of crisis is inevitable. Hence attention is being diverted towards the use of biofertilizing agents from crop production; this is not a new area. Researchers have indicated the plant products as a great potential in the use for crop productivity. Hence while evaluating them for the nematode control observations were also recorded on the mycorrhizal infection and rhizobial nodulation; and weight of nodule; both the agents were favoured by the neem based toxicants.

This is an additional premium received by these botanical pesticides which are environmentally and ecologically safe and sound and can be incorporated in the integrated control of pest and disease control (Table-4). There are reports that application of mycorrhizae in soil also control nematodes. Hence a separate study was undertaken to eliminate this. However, the control obtained of nematodes in this study can not be ruled out the possibilities of role of mycorrhizae.

Table 4. Effect of soil application of neem cake, neem oil, Repellin and carbofuran at the rate of 1 and 2 per cent of soil weight on rot, nodule number and nodule weight, plant height, plant dry weight, seed weight for plant and yield of soybean (cultivar - JS-72-44)

Treatment	Conc. (%)	Germination (%)	Nodule		Shoot		Root		Yield
			Number	Weight	Length (cm)	Dry Weight (g)	Length (cm)	Dry Weight (g)	per hactare (quint)
Neem cake	1.00	70 b	80 b	10.00 b	36.91 c	10.91 b	13.00 b	2.64 c	15.00 c
	2.00	75 a	85 a	18.00 b	38.91 c	11.92 b	14.00 b	3.62 c	16.00 c
Neem oil	1.00	60 c	70 c	8.00 c	47.43 b	12.91 b	14.07 b	2.86 c	15.50 c
	2.00	65 b	75 b	9.00 b	49.23 b	12.00 b	14.50 b	3.50 c	15.50 c
Repellin	1.00	70 b	85 a	10.00 b	51.99 b	13.58 b	17.81 a	5.63 a	22.00 a
	2.00	75 a	90 a	12.00 b	57.80 a	17.81 a	19.82 a	5.83 a	23.00 a
Carbofuran	1.00	65 b	60 c	7.00 c	49.00 b	11.68 b	14.07 b	4.64 b	18.00 b
	2.00	65 b	65 c	8.00 c	50.00 b	12.00 b	25.00	4.90 b	18.00 b
Check	—	35 d	05 c	4.00 c	29.00 d	08.80 c	10.75 c	2.64 c	12.00 d

* Data are mean of three replications

** Seed treatment at the rate of 0.2 mg a.i. per kg was given with thiram and soil inoculation with mycorrhizal fungi and seed inoculation with *Rhizobium japanicum*; before sowing;

*** Number not followed by the same letter are significantly different at the P=0.05 in Duncan's multiple range tests;

**** Soil was inoculated with *Sclerotium rolfsii*, and *Rhizoctonia bataticola* before application of the treatment

***** Disease index after 40 days after sowing

The objectives of the pest control also envisages increase in the yield of the crop to treated. The application of neem based pesticides increased the yield contributing factors and yield of soybean significantly in the present investigation. The data pertaining to nodule number and weight, germination percentage, plant height and dry weight and grain yield were significantly increased over the control. The increase was reverse of toxicity of the pesticides; here the most toxic pesticides increased more yield than others. Because of the reduction in the nematode population and diseases index or intensity (Table-4). Such increase has been reported in maize (Egunjobi and Afoloni, 1976) and bhindi (Kathirvel *et al.* 1992; Mammen, 1972).

One of the interesting findings of the present investigation is that the neem based products are fungicidal as well as nematicidal. Regarding the mechanism of control of nematodes by application of neem cake, oil and Repellin evidences suggest that they encountered many saprophytic soil biota which parasitised the nematodes and possess antifeedants and hatching activities (Bhatnagar *et al.*, 1980). Further they observed that the neem contains azairachtin which is highly lethal to the larvae of *Meloidogyne incognita*. The neem leaf extract has been demonstrated to inhibit hatching of nematodes eggs (Hussain and Mazoodm 1976 and Ayoub, 1977). The reduction of soilborne plant pathogens was also observed which may be due to competition for space and food; production of antibiotic substances and direct penetration by saprophytic fungi (Vyas, 1993).

Neem based products such as cakes, oil and leaf extracts are being used intensively for the control of insect-pest and diseases in agriculture and domestic uses under different trade names and formulations in several countries. The results of this control are very encouraging. hence the efficacies of the method developed at this centre has been demonstrated in field research trials and on farmer's fields. The treatments used in the experiments were simulated on the farmer's field and significant increase in the yields were observed and the results were presented in the Table-5 which are self explanatory.

Table 5. Resultsl of large plot research farm demonstrations on control of root rot complex and root knot of soybean using neem cake and Repellin on the farmer's fields which have history of soybean cultivation since last twenty years

Details of Treatment	Plant stand ha^{-1}	Increase in plant stand (%)	Yield ($Q\ ha^{-1}$	Increase in yield (%)	Gross Earning (Rs)
Farmer A					
Neem cake (1 %)	4,00,000	235.29	25.00	250	20,000.00
Repellin (1%)	3,75,000	220.58	22.00	220	17,600.00
Carbofuran (1%)	3,60,000	211.76	21.50	215	16,800.00
Check	1,70,000		10.00	—	
Farmer B					
Neem cake (1%)	4,00,000	222.22	22.50	225	18,000.00
Repellin (1%)	4,00,000	222.22	22.00	220	17,600.00
Carbofuran (1%)	3,80,00	211.11	16.00	145	12,800.00
Check	1,80,000		11.00	—	

* All the treatments were simulated as per experimental results presented in the tables.

This seems to be first report where disease significant reduction soybean particulary root rot and collar rot and also nematode population has been reported and also significant increase in yield was recorded.

REFERENCES

Ayoub, S.M. 1977. Plant Nematology - An Agricultural Training Aid. Department of Food and Agriculture, Division of Plant Industry, Laboratory Services - Nematology, California, 157 pp.

Bhatnagar, A.K., T.K. Mukerjee and B. Tyagi. 1980. Effect of leaves of some of plants on Meloidogyne incognita. *Andhra Agric. J.,* **27**: 312-314.

Christie, J.R. and V.G. Perry 1952. Removing Nematodes from Soil. *Proc. Helminthol. Soc. Wash.,* **18**: 9-13.

Devkumar, C., B.K. Goswami and S.K. Mukheree. 1985. Nematicidal principal from neem screening of neem kernel fractions against *Meloidogyne incognita. Neem Letter,* **2**(2): 21.

Egunjobi, O.A. and S.O. Afolani. 1976. Effect of neem leaf extracts on populations of *Pratylenchus brachyurus* and on growth and yield of maize. *Nematologiá,* **22**: 125-132.

Hussain, S.I. and A. Maszood. 1976. Effect of some Plant extracts on larval hatching of *Meloidogyne incognita. Acta. Bot. India,* **3**: 142-146.

Kathirvel, M., G. Balasubramanian, M. Gopal and C.V. Sivakumar. 1992. Effect of seed treatment with botanicals and chemicals for the control of root knot nematodes, *Meloidogyne incognita* infesting okra, *Abelmoschus esculentus L. Indian J. Plant Prot.,* **20**: 191-194.

Mail, W.F. and H.H. Lyon. 1975. Pictorial key to Genera of Plant Parastic Nematodes. Comstock Publ. Assoc. Ithica, New York, 219 pp.

Mammen, K.V. 1972. Effect of oil cakes on the incidence of root galls and the yield of *bhindi* in nematode infested soil. *Agric. Res. J. Kerala,* **10**: 186-187.

Murthy, M.M.K. and K.T. Rao. 1992. Control of root knot nematode *Meloidogyne incognito* on betalvine with Gertom oil cake amendment and now volatile nematodes, *Indian J.Pl. Prot.,* **20**: 171-173.

Swaroop, Gopal and D.R. Dasgupta (ed). 1986. Plant Parasitic nematode of India and progress. Indian Agricultural Research Institute, New Delhi, 497 pp.

Taylor, A.L. 1967. Introduction to Research on Plant Nematology. An FAO Guide to the study and control of plant parasitic nematodes, FAO, Rome 133 pp.

Vyas, S.C. 1993. Annual Report on "Studies on biological control of soilborne pathogens of wheat and soybean. M.P. Council of Science and Technology, Bhopal, J.N. Agri. Univ. Indore, 110 pp.

Neem for the Management of Crop Diseases, 1995
Editor : V. Mariappan
Associated Publishing Co., New Delhi, India
pp. 209-211.

CONTROL OF ROOT-KNOT NEMATODE WITH NEEM PRODUCTS

Mashkoor Alam, M., Parveen, Ghazala and Khan, Shahabuddin

Department of Botany, Aligarh Muslim University, Aligarh-202002, India

Neem (*Azadirachta indica* A. Juss.; family-Meliaceae) has become cynosure of plant protection scientists engaged in the control of plant pests and diseases. Alam (1993) has reviewed the work on bio-activity of neem against plant-parasitic nematodes. Various neem products, e.g., seed cake, plant parts/extracts, wood saw dust and industrial products are being used incessantly in such studies. 'Nimin' and plant oils including neem oil as urea coating agents have been reported to be effective in suppressing the activity of plant-parasitic nematodes attacking tomato and chilli resulting in improved plant growth (Akhtar and Alam, 1993). The present investigation was undertaken to evaluate the efficacy of 'Nimin' and plant products of neem and a related species, *Melia azedarach* L. (Chinaberry/rocket salad/Bakain) as seed dressing agents for the control of root-knot nematode attacking okra.

MATERIALS AND METHODS

Seeds of okra (*Abelmoschus esculentus* Moench.) cv. Seven Dhari *were treated with 'Nimin'*[a], neem oil[b], slurry[c] of neem cake, and fine paste[d] of neem and bakain leaves and bakain fruits separately so as to get a uniform coating of the materials on the seeds, which were then dried under shade. Five seeds for each treatment were sown in 15 cm diameter earthen pots each containing 1 kg autoclaved soil. After seed germination, thinning was done to retain a single plant per pot. Plants were inoculated with 5000 freshly hatched second stage juveniles (J2) of the root-knot nematode, *Meloidogyne incognita* three weeks after the emergence of seedling. Untreated pots served as control. Each treatment was replicated three times. The experiment was terminated three months after nematode inoculation. Plant growth parameters and root-knot index were also determined. The root-knot index was recorded on a 0-5 scale (Sasser *et al.*, 1984).

[a] 'Nimin' is a triterpene rich neem-based product of Godrej Soaps Ltd., Bombay, India.

[b] Neem oil was obtained from an agro-industrial expeller plant where oil is extracted from dried fruits of neem by expression method.

[c] Neem cake is the solid residue left after extraction of oil from dried fruits. It is generally used as organic manure (containing about 5% N) by farmers. Twenty five gram neem cake was soaked in 25 ml distilled water for overnight and then homogenised in a warring blender to obtain the slurry.

[d] Fresh leaves/fruits (25 g each) were separately crushed in a mortar and pistle by adding 5 ml distilled water so as to obtain a fine paste.

RESULTS

It is clear from the results presented in Table 1 that seed dressing treatment with different products of neem and Chinaberry significantly inhibited the root-knot development on okra caused by *Meloidogyne incognita*. The root-knot index was as low as 1.1 and 1.8 in plants raised from neem oil and 'nimin' treated seeds, respectively, as compared to 3.6 in control plants. In the remaining treatments, the root-knot indices ranged between 2.0 and 2.6 which were significantly lower than the untreated control.

As a result of suppression in the root-knot incidence, plant growth (length as well as weight of shoot and root) improved. Infection of plants with the root-knot nematode brought about reduction in plant weight from 18.5 g to 9.7 g (Table 1). However, there was a significant improvement in plant weight in different seed dressing treatments.

It was interesting to note that in case of neem oil treated seeds, plant weight in the inoculated set was even more than the untreated-uninoculated control, thereby, indicating some stimulating effect of the treatment on the plant growth (Table 1). A similar trend was also noted in case of plant length in the same treatment (Table 1).

Table 1. Effect of seed dressing with natural products on the root-knot development and plant growth of okra (Values are mean of three replicates).

Treatment	Mean length (cm)			Mean weight (g)			Mean root-knot index (0-5 scale)
	Shoot	Root	Total	Shoot	Root	Total	
Untreated (uninoculated)	36.0	30.0	66.0	12.9	5.6	18.5	—
Untreated (inoculated)	31.6	25.0	56.6	7.6	2.1	9.7	3.6
Nimin (")	46.3	31.0	77.3	10.8	4.7	15.5	1.8
Neem oil (")	52.6	40.0	92.6	14.1	6.2	20.3	1.1
Neem cake (")	37.0	30.3	67.3	12.1	4.5	16.6	2.6
Neem leaf (")	49.3	28.0	77.3	12.6	4.1	16.7	2.0
Bakain leaf (")	44.3	33.3	77.6	13.5	4.1	17.6	2.3
Bakain fruits (")	46.0	33.3	79.3	13.3	4.9	18.2	2.6
C.D. (P=0.05)	7.2	6.2		1.6	0.9		0.6
C.D. (P=0.01)	10.0	8.7		2.2	1.3		0.8

DISCUSSION

The different seed dressing materials used in the present study, viz., 'nimin' and various neem and chinaberry products brought about a significant control of root-knot nematode, *Meloidogyne incognita* on okra. This is understandable as the test materials are rich in bio-active ingredients including azadirachtin, an important triterpenoid, which might be responsible for the changes in the microclimate around the seed-coat rendering it unfavourable for nematode activity (Alam, 1993). These results are in consonance with the findings of Vijayalakshmi and Goswami (1986), Dash and Padhi (1990), Khan and Husain (1988) and Siddiqui and Alam (1987). Siddiqui and Alam (1988, 1990) obtained significant control of *M. incognita* on tomato, egg plant and okra by treating the seeds with azadirachtin and nimbin.

The test materials might have also induced some resistance in the emerging roots against the attack of root-knot nematode (Alam, 1993; Singh *et al.*, 1980).

Stimulated plant growth in one of the treatments might have been due to its influence on the metabolism of germinating seeds (Siddiqui and Alam, 1988).

REFERENCES

Akhtar, M. and M.M. Alam. 1993. Control of plant-parasitic nematodes by 'Nimin'—an urea-coating agent and some plant oils. *Zeitschrift fur Pflanzenkrankheiten und Pflanzenschutz.*, **100**: 337-342.

Alam, M.M. 1993. Bio-activity against phytonematodes. In *Neem Research and Development* (Eds.- Randhawa, N.S. and Parmar, B.S.), pp. 123-143, Society of Pesticide Science, India, New Delhi.

Dash, B.K. and N.N. Padhi. 1990. Response of root-knot nematode infecting tomato to seed treatment by some neem products. *Orissa J. Agr. Res.*, **2**: 148-150.

Khan, T.A. and S.I. Husain. 1988. Studies on the efficacy of seed treatment with pesticides, oil-cakes, neem leaf and culture filtrate of *Paecilomyces lilacinus* for the control of diseases caused by the presence of *Rotylenchulus reniformis, Meloidogyne incognita* and *Rhizoctonia solani* either individually or concomitantly on cowpea. *Indian J. Nematol.*, **18**: 192-198.

Sasser, J.N., C.C. Carter, and K.M. Hartman. 1984. *Standardization of Host Suitability Studies and Reporting of Resistance to Root-knot Nematodes.* CNRCP Publication, Department of Plant Pathology, N.C. State University and U.S.A.I.D., Raleigh, N.C. pp. 7.

Siddiqui, M.A. and M.M. Alam. 1987. Efficacy of seed dressing with extracts of neem and Persian lilac against *Meloidogyne incognita* and *Rotylenchulus reniformis. Nematol. medit.*, **15**: 399-403.

Siddiqui, M.A. and M.M. Alam. 1988. Effect of seed treatment with azadirachtin on root-knot development on, and growth of some vegetables. *Tests Agrochem. Cult. 9 Ann. appl. Biol.*, **112**(Suppl.): 20-21.

Siddiqui, M.A. and M.M. Alam. 1990. Control of root-knot, reniform and stunt nematodes with nimbin seed dressing. *Nematol. medit.*, **18**: 19-22.

Singh, S.P., M. Ahmad, A.M. Khan and S.K. Saxena. 1980. Effect of seed treatments with certain oil-cakes or nematicides on the growth of tomato and on rhizosphere population of nematodes and fungi. *Nematol. medit.*, **8**: 193-198.

Vijayalakshmi, K. and B.K. Goswami. 1986. Effect of seed treatments with neem cake and neem oil on the germination of moong (green gram) and its vulnerability to root-knot nematodes. *Int. Nematol. Network Newsl.*, **3**(1): 8-9.

Neem for the Management of Crop Diseases, 1995
Editor : V. Mariappan
Associated Publishing Co., New Delhi, India
pp. 213-220.

INTEGRATED CONTROL OF PLANT PARASITIC NEMATODES WITH OIL-SEED CAKES/NEMATICIDES ON CARROT

Siddiqui, Mansoor A. and Mashkoor Alam, M.

Department of Botany, Aligarh Muslim University, Aligarh-202 002, India

It has been largely realized that pesticides available today are generally expensive and hazardous in nature. This has triggered intensive research to explore possibilities of evolving alternative methods. In case of plant parasitic nematodes, organic amendments, particularly oil-seed cakes have shown great promise in their management (Alam, 1990). Neem cake and plant parts have been claimed as one of the most efficacious materials for combating nematode menace (Siddiqui, 1986; Alam, 1993). In the present investigation an attempt has been made to evaluate the efficacy of neem and other cakes and nematicides against plant parasitic nematodes infesting carrot under two tillage regimes.

MATERIALS AND METHODS

A field harbouring high population of plant parasitic nematodes viz., *Hoplolaimus indicus* Sher, *Helicotylenchus indicus* Siddiqi, *Rotylenchulus reniformis* Linford & Oliveira, *Tylenchorhynchus brassicae* Siddiqi, *Tylenchus filiformis* Butschli and *Meloidogyne incognita* (Kofoid & White) Chitwood was selected and thoroughly prepared. The field was divided into two portions receiving normal (20 cm deep) and deep (40 cm deep) ploughing treatments respectively. Both the sub-plots were divided into beds of 9 m^2 size each separated by a 0.5 m wide buffer zone. The beds were given the following treatments: inorganic fertilizers, organic amendments and nematicides. Untreated beds served as control. In the case of inorganic fertilizers, urea @ 110 kg N/ha, superphosphate @ 55 kg P/ha and murate of potash @ 55 kg K/ha were incorporated into the soil. In the case of organic amendments, compost and oil-seed cakes of margosa/neem (*Azadirachta indica* A. Juss.), castor (*Ricinus communis* L.), mustard (*Brassica campestris* L.), rocket-salad/Duan (*Eruca sativa* Mill.) and groundnut (*Arachis hypogaea* L.) were added to the soil @ 110 kg N/ha. In the last category two nematicides viz., carbofuran/furadan-3G (2,3-dihydro-2,2-dimethyl-7-benzofuranyl methyl carbamate) and aldicarb/Temik-10G (2-methyl-2-(methylthio) propionaldehyde-0-(methylcarbamoyl) oxime) were added to the soil @ 1 kg a.i./ha. There were 3 replicates for each treatment arranged in randomized block design (Fig. 1; Panse and Sukhatme, 1978). Immediately after the soil treatment, the beds were watered for ensuring proper decomposition of the organic addities. Seeds of carrot (*Daucus carota* L.) cv. 'Pusa Kesar' were sown after a waiting period of two weeks. Weeding, watering etc. were done as and when considered necessary.

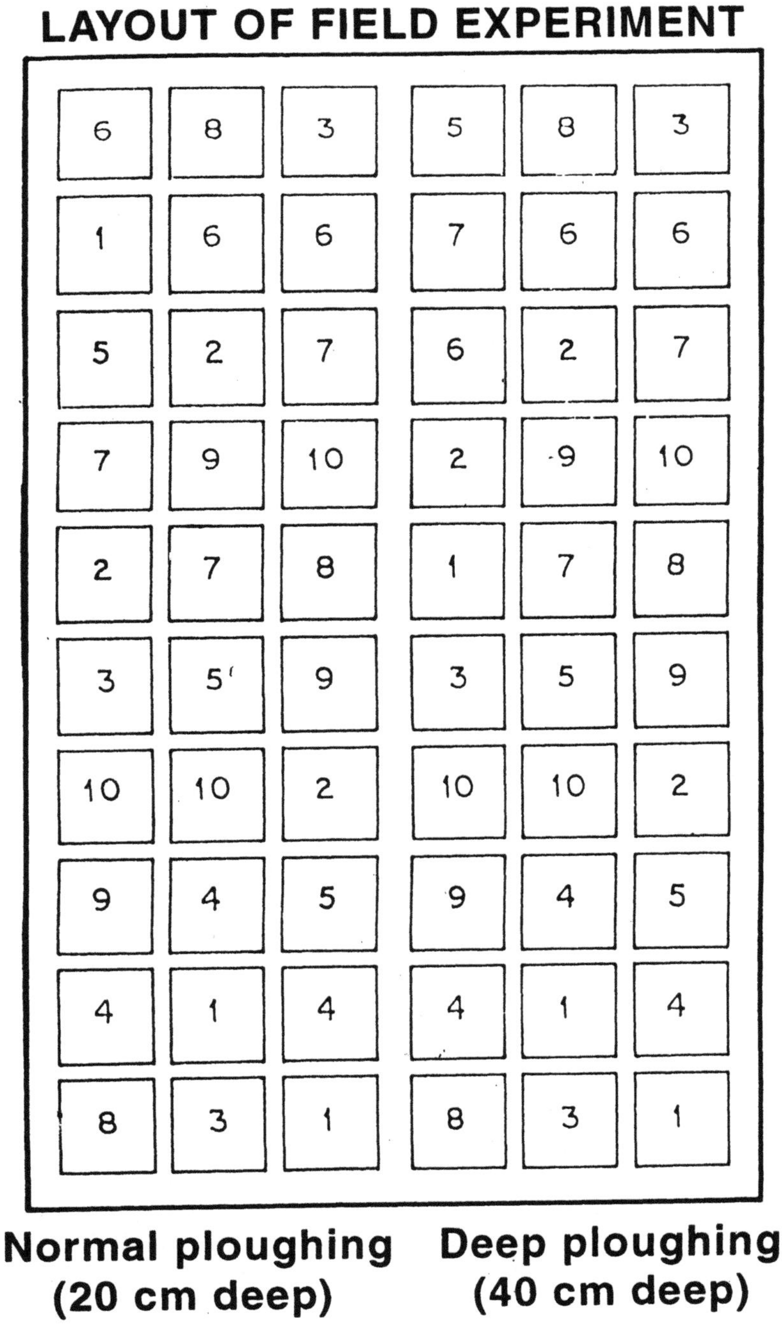

Fig. 1. Layout of the field experiment showing randomised arrangement of different treatments.

Initial (before treatment) and final (3 months after sowing) nematode populations were determined at the time of sowing and termination of the experiment from each bed following Cobb's sieving and decanting and Baermann's funnel techniques (Southey, 1986). Yield was also noted.

For studying the residual effect of different treatments, the beds of the first experiment were ploughed (normal and deep were repeated) just after the completion of the first experiment and half dozes of inorganic fertilizers i.e., urea @ 55 kg N/ha, superphosphate and muriate of potash @ 27.50 kg P and K/ha were incorporated in all the beds. No waiting period was given in this experiment and three-week-old seedlings of tomato (*Lycopersicon lycopersicum* (L.) Karsten) were transplanted soon after the application of inorganic fertilizers. Aftercare of the experiment such as weeding and watering was done when necessary.

Nematode populations were determined three months after transplantation in the same manner as described above and plant growth parameters (fresh weights of shoot and root) noted. The data were statistically analysed for critical difference (C.D.) at $P = 0.05$ and $P = 0.01$ (Panse and Sukhatme, 1978)

RESULTS

The results summarized and presented in Tables 1 and 2 clearly indicate that the populations of plant parasitic nematodes were significantly reduced in beds given different treatments. However, there was manifold increase in nematode populations in untreated control. Margosa/neem, rocket-salad/Duan, mustard and groundnut cakes were significantly more effective than the inorganic fertilizers and compost. But they were slightly less effective than the two test nematicides viz., aldicarb and carbofuran. Deep ploughing (40 cm deep) of soil also brought about significant reduction in nematode populations. The efficacy of deep ploughing (40 cm deep) enhanced when it was combined with oil-cakes/nematicides (Table 1). The results of the second experiment (Table 2) clearly show that the oil-cakes/nematicides remained effective in soil against the nematode populations even in the subsequent crop, tomato. The population of *M. incognita*, *R. reniformis* and *T. brassicae* remained high throughout the experiment in untreated beds which shows that the test crops were good hosts to these nematode species. The number of other nematodes viz., *Hop. indicus*, *Hel. indicus* and *T. filiformis* remained comparatively low.

Highest yield of carrot was observed in the beds treated with aldicarb both in normal (500 q/ha) as well as in deep (525 q/ha) ploughed beds. Among oil-cakes, the yield was significantly enhanced in all the cases. The combination of oil-cakes and nematicides with deep ploughing enhanced the yield (Table 1). The plant growth (fresh weights of shoot and root) of tomato was significantly high when it was grown in beds treated with oil-cakes and nematicides in the preceding crop. Oil-cakes and nematicides integrated with deep ploughing gave much healthier plants. Nematicides were found to be slightly more beneficial in promoting the plant growth than

Table 1. Combined effect of organic soil amendments/nematicides and ploughing (normal/deep) on the population of plant parasitic nematodes and yield of carrot cv. Pusa Kesar in field.

Treatment	Population of plant parasitic nematodes per 200 g soil																Yield/ha (q)	
	Normal ploughing (20 cm deep)								Deep ploughing (40 cm deep)								Normal ploughing	Deep ploughing
	Hop	Hel	Rot	Trh	Tyl	Mel	Total	Sap	Hop	Hel	Rot	Trh	Tyl	Mel	Total	Sap		
Untreated	150	70	320	150	100	1100	1890	983	130	70	287	150	50	1088	1775	700	233.33	263.25
Inorg. Fert.	20	40	337	96	40	896	1429	486	23	85	228	142	43	749	1270	1176	315.00	321.16
Compost	27	30	378	130	93	900	1558	890	30	43	275	109	50	859	1366	456	275.00	287.00
Castor cake	30	49	203	119	23	466	890	780	23	47	138	100	23	242	573	400	377.67	415.50
Neem cake	30	59	105	71	20	280	565	730	17	36	80	41	23	261	458	563	435.00	455.00
Mustard cake	23	43	227	83	27	349	752	496	40	60	118	79	23	284	604	366	413.75	437.50
Duan cake	20	50	106	26	26	126	353	383	33	74	120	79	20	182	508	280	425.00	475.00
Groundnut cake	47	37	110	66	40	235	535	666	57	40	86	53	43	169	448	456	452.50	487.50
Carbofuran + Inorg. Fert.	53	20	91	33	20	183	400	413	23	30	79	67	20	128	347	400	471.67	501.83
Aldicarb + Inorg. Fert.	20	20	33	76	47	146	342	536	33	20	73	40	20	120	306	530	500.00	525.00
Initial population	147	116	340	132	120	360	1215	1300	147	116	340	132	120	360	1215	1300		
C.D. (P = 0.05)							53.69								47.94		78.74	102.17
C.D. (P = 0.01)							75.39								65.70		107.87	139.96

Each value is an average of three replicates.

Hop = *Hoplolaimus indicus*, Hel = *Helicotylenchus indicus*, Rot = *Rotylenchulus reniformis*, Trh = *Tylenchorhynchus brassicae*, Tyl = *Tylenchus filiformis*, Mel = *Meloidogyne incognita* larvae, Total = Total Tylenchids, Sap = Total saprozoic.

the oil-cakes (Table 2). It appears from the results that reduction in nematode population due to different treatments has positive correlation with the improvement in plant growth.

DISCUSSION

It is evident from the results that the nematodes commonly inhabiting the soil during the field trials viz., *M. incognita, R. reniformis, T. brassicae, Hop. indicus, Hel. indicus and T. filiformis* were significantly reduced by oil-seed cakes of margosa/neem, mustard, rocket-salad/*Duan* and groundnut. Due to the reduction in nematode populations, the yield of carrot enhanced. Our findings with respect to the oil-seed cakes are in general agreement with those of Singh and Sitaramaiah (1971) and Alam (1989). The effect of *duan* cake was reported for the first time.

In the present findings oil-seed cakes were found effective in reducing the nematode population even in the subsequent crop (Table 2). It may be due to the slow decomposition of oil-cakes releasing nemato-toxic substances for a longer time. Our results are in conformity with Singh and Sitaramaiah (1966) and Alam (1990).

Though it is difficult to pin point the exact mechanism involved in the reduction of nematode population by the application of oil-seed cakes to the soil yet many scientists have put forward different theories in this regard. Alam *et al.* (1979), after extensive work on the mechanism of oil-seed cakes pointed out that oil-seed cakes undergo decomposition in moist soil and release many compounds viz., phenols, aldehydes and different gases, including ammonia. Fatty acid was also released after the decomposition of oil-seed cakes (Sitaramaiah and Singh, 1978). Under laboratory conditions all these chemicals were found to be highly nemato-toxic (Walker *et al.*, 1967; Alam *et al.,* 1979). Some neem chemicals were found effective in reducing the nematode population (Siddiqui and Alam, 1988, 1989). Water extracts of oiled and deoiled cakes have also been found detrimental to many genera of nematodes (Alam *et al.*, 1982). They also check juvenile hatching because they contain toxic chemicals (Khan *et al.*, 1974; Alam *et al.*, 1979).

The decomposition of oil-cakes takes place due to the increased activity of soil microorganisms including bacteria, fungi etc. (Kirmani, 1977), This may also lead to an increased predacious and parasitic activity of soil biota. Secondly, the organic additives usually change physical and chemical conditions of the soil (Ahmad *et al.*, 1972) and improve soil conditions for greater root growth thereby enhancing the utilization of soil nutrients; as a consequence the nematode damage is masked or minimized and the host-nematode relationship is altered. Van der Laan (1956) claimed that due to organic amendments some physiological changes take place in the host plant which render them unfavourable to the nematode development. Similarly, in the case of oil-cakes this was proved by Alam *et al.* (1980) and Sitaramaiah and Singh (1978) who confirmed an increased resistance in plants grown in soil amended with oil-cakes. They suggested that this induction of resistance was due to the increased level of phenolics in the host roots. It was reported that the

Table 2. Residual effect of organic soil amendments/nematicides and ploughing (normal/deep) on the population of plant parasitic nematodes and plant growth of tomato cv. Pusa Ruby in field. (Crop in the preceding season was carrot cv. Pusa Kesar).

Treatment	Population of plant parasitic nematodes per 200 g soil																Weight (g)					
	Normal ploughing (20 cm deep)								Deep ploughing (40 cm deep)								Normal ploughing			Deep ploughing		
	Hop	Hel	Rot	Trh	Tyl	Mel	Total	Sap	Hop	Hel	Rot	Trh	Tyl	Mel	Total	Sap	Shoot	Root	Total	Shoot	Root	Total
Untreated	53	93	366	300	78	1336	2226	1463	43	100	400	212	80	1115	1950	1296	20.17	1.50	21.67	28.00	2.33	30.33
Inorg. Fert.	117	122	273	133	37	887	1569	1460	77	70	178	111	46	771	1253	1416	27.83	2.83	30.66	35.33	4.33	39.66
Compost	93	116	400	145	104	1084	1942	1353	47	77	390	89	80	1044	1727	1366	31.83	6.83	36.66	36.27	4.16	40.43
Castor cake	33	40	331	92	97	376	895	920	67	98	132	93	57	196	643	933	30.00	6.00	36.00	39.00	9.00	48.00
Neem cake	47	160	116	83	27	183	516	783	40	47	132	53	23	224	519	713	28.83	3.42	32.25	36.33	6.17	42.50
Mustard cake	70	49	211	79	27	347	783	766	46	53	167	78	40	226	610	716	29.83	4.83	34.66	43.83	4.83	48.66
Duan cake	33	60	200	60	23	126	502	626	27	40	101	93	44	159	464	983	31.83	4.30	36.13	39.33	5.33	44.66
Groundnut cake	43	53	156	63	40	280	630	800	20	43	141	77	38	174	493	616	38.00	5.33	43.33	52.00	3.20	50.10
Carbofuran + Inorg. Fert.	53	77	115	94	23	171	533	316	23	37	110	47	33	122	372	296	40.17	5.08	45.25	50.83	8.47	59.30
Aldicarb + Inorg. Fert.	45	50	136	40	20	194	475	370	20	33	62	30	33	109	287	283	40.30	4.26	44.56	54.00	7.00	61.00
C.D. (P = 0.05)							479.30								483.10				20.18			17.50
C.D. (P = 0.01)							656.70								661.80				27.55			23.97

Each value is an average of three replicates.
Hop = *Hoplolaimus indicus*, Hel = *Helicotylenchus indicus*, Rot = *Rotylenchulus reniformis*, Trh = *Tylenchorhynchus brassicae*, Tyl = *Tylenchus filiformis*, Mel = *Meloidogyne incognita* larvae, Total = Total Tylenchids, Sap = Total saprozoic.

penetration of second stage juveniles and the subsequent root galling was greatly inhibited in seedlings raised in oil-cake amended soil.

The plant parasitic nematodes were also greatly influenced by the depth of ploughing. It was found during the field trial that the deep ploughing (40 cm deep) brought about significant reduction in the population of plant parasitic nematodes. Oil-seed cakes/nematicides integrated with deep ploughing were found to be more efficacious than the oil-seed cakes/nematicides applied in normal ploughed beds. The suppressive effect of deep ploughing has been reported by many workers (Thomas, 1978; Caveness, 1979; Khan and Saxena, 1980). It has been suggested that the deep ploughing distrubs the ecological set up of nematodes which are exposed to external unfavourable conditions and thus their populations decline (Khan and Saxena, 1980).

REFERENCES

Ahmad, R., A.M. Khan, and S.K. Saxena, 1972. Changes resulting from amending the soil with oil-cakes and analysis of oil-cakes (Abstr.). *Proc. 59th Ses. Indian Sci. Cong.* Part II, 164 pp.

Alam, M.M. 1989. Control of root-knot and stunt nematodes with horn meal, bone meal and oil-seed cakes. *Indian J. Nematol.*, **19**: 166-170.

Alam, M.M. 1990. Control of plant parasitic nematodes with organic soil amendments and nematicides in nurseries of annual plants. *J. Bangladesh Acad. Sci.*, **14**: 107-113.

Alam, M.M. 1993. Bioactivity against phytonematodes, In: *Neem Research and Development* (Eds. N.S. Randhawa and B.S. Parmar). Soc. Pesticide Sci. India Publication No. 3. pp. 123-143.

Alam, M.M., M. Ahmad and A.M. Khan, 1980. Effect of organic amendments on the growth and chemical composition of tomato, *Meloidogyne incognita*. *Plant and Soil*, **57**: 231-236.

Alam, M.M., A.M. Khan and S.K. Saxena, 1979. Mechanism of control of plant parasitic nematodes as a result of the application of organic amendments to the soil. V. Role of phenolic compounds. *Indian J. Nematol.*, **9**: 136-142

Alam, M.M., A.M. Khan and S.K. Saxena, 1982. Relative toxicity of decomposed and undecomposed oil-cakes to plant parasitic nematodes. *Acta Bot. Indica.*, **10**: 124-127.

Caveness, F.E. 1979. Nematode populations under a no-tillage soil management regime. In: *Soil tillage and crop production* (Proc. Series No. 2 Int. Inst. Trop. Agr., Ed. Lal, R.). Ibadan, Nigeria: 135-145.

Khan, A.M., M.M. Alam and R. Ahmad 1974. Mechanism of the control of plant parasitic nematodes as a result of the application of oilcakes to the soil. *Indian J. Nematol.*, **4**: 93-96.

Khan, A.M. and S.K. Saxena, 1980. Effect of certain farm practices on population of nematodes. *J. Environ. Res.*, **1**: 12-16.

Kirmani, M.R. 1977. Studies on the role of fungi from decomposed oil-cakes on root-knot development on eggplant, *Solanum melongena*. Ph.D. Thesis, Department of Botany, A.M.U., Aligarh, 145 pp.

Panse, V.G. and Sukhatme, P.V. 1978. Statistical methods for agricultural workers (Revised by P.V. Sukhatme and V.N. Amble), I.C.A.R., New Delhi, 347 pp.

Siddiqui, M.A. 1986. Studies on the effect of organic soil amendments on plant parasitic nematodes. Ph.D. Thesis, Department of Botany, A.M.U., Aligarh, 402 pp.

Siddiqui, M.A. and M.M. Alam, 1988. Effect of seed treatment with azadirachtin on root-knot development on, and growth of some vegetables. *Tests of Agrochemicals and cultivars,* **9** (Annals of Applied Biology 112 Supplement) England: 20-21.

Siddiqui, M A. and M.M. Alam, 1989. Efficacy of azadirhtin as seed treatment for the control of the reniform nematode on some vegetables. *Tests of Agrochemicals and cultivars* **10** (Annals of Applied Biology 114 Supplement) England: 20-21.

Singh, R.S. and K. Sitaramaiah, 1966. Incidence of root-knot of okra and tomatoes in oilcakes amended soil. *Plant Dis. Reptr.,* **50**: 668-672.

Singh, R.S. and K. Sitaramaiah, 1971. Contról of root-knot through organic and inorganic amendments of soil: Effect of oilcakes and sawdust. *Indian J. Mycol. Pl. Pathol.,* **1**: 20-29.

Sitaramaiah, K. and R.S. Singh, 1978. Role of fatty acids in margosa cake applied as soil amendment in the control of nematodes. *Indian J. Agr. Sci.,* **48**: 266-270.

Southey, J.F. 1986. Laboratory methods for work with plant and soil nematodes. Min. Agr. Fish. Food, HMSO, London.

Thomas, S.H. 1978. Population densities of nematodes under seven tillage regimes. *J. Nematol.,* **10**: 24-27.

Van der Laan, P.A. 1956. The influence of organic manuring on the development of the potato eelworm, *Heterodera rostochiensis. Nematologica.,* **1**: 112-125.

Walker, J.T., C.H. Specht and S. Mavrodineau 1967. Reduction of lesion nematode in soybean meal and oil amended soil. *Plant Dis. Reptr.,* **51**: 1021-1024.

www.ingramcontent.com/pod-product-compliance
Ingram Content Group UK Ltd.
Pitfield, Milton Keynes, MK11 3LW, UK
UKHW020811290926
15470UKWH00003B/1082

9 789351 309840